PROGRESS IN THE NEUROSCIENCES AND RELATED FIELDS

Studies in the Natural Sciences

A Series from the Center for Theoretical Studies
University of Miami, Coral Gables, Florida

ORBIS SCIENTIAE

PROGRESS IN THE NEUROSCIENCES AND RELATED FIELDS

Chairman
Behram Kursunoglu

Editors
Stephan L. Mintz
Susan M. Widmayer

Scientific Secretaries
Chui-Shuen Hui
Joseph Hubbard
Joseph F. Malerba

Center for Theoretical Studies
University of Miami
Coral Gables, Florida

PLENUM PRESS • NEW YORK AND LONDON

Library of Congress Cataloging in Publication Data

Orbis Scientiae, University of Miami, 1974.
Progress in the neurosciences and related fields.

(Studies in the natural sciences, v. 6)
"Proceedings of Orbis Scientiae held by the Center for Theoretical Studies, University of Miami, January 7-11, 1974."
Includes bibliographical references.
1. Neurophysiology—Congresses. I. Kurşunoğlu, Behram, 1922- II. Mintz, Stephan, ed. III. Widmayer, Susan M., ed. IV. Miami, University of, Coral Gables, Fla. Center for Theoretical Studies. V. Title. VI. Series. [DNLM: 1. Biochemistry—Congresses. 2. Biophysics—Congresses. 3. Neurophysiology—Congresses. W30R49 1974p / WL102 064 1974p]
QP351.07 1974 612'.82 74-10822
ISBN 0-306-36906-0

Part of the Proceedings of Orbis Scientiae held by the Center for Theoretical Studies, University of Miami, January 7-11, 1974

A Division of Plenum Publishing Corporation
227 West 17th Street, New York, N.Y. 10011

United Kingdom edition published by Plenum Press, London
A Division of Plenum Publishing Company, Ltd.
4a Lower John Street, London, W1R 3PD, England

Printed in the United States of America

Preface

This volume contains the papers presented during the Neurophysiology Session of the first Orbis Scientiae of the Center for Theoretical Studies, University of Miami, Coral Gables, Florida. With this first Orbis which met from January 7 - 11, 1974, the Center for Theoretical Studies has inaugerated a new set of annual gatherings devoted to the natural sciences and to problems on the "interface" of science and society.

The content of the talks presented ranged over a wide variety of topics in neurophysiology, biophysics and biochemistry. A number of the talks concerned various aspects of the brain and its functions. Recent results were also presented on the physics and chemistry of membranes. The papers in the volume are presented in the order in which they were originally delivered.

Special thanks are due to Mrs. Helga Billings, Miss Sara Lesser and Mrs. Jacquelyn Zagursky for the typing of the manuscript and for their efficient and cheerful attention to the details of the conference.

THE EDITORS

Contents

From Left: Sir John Eccles, Lady Eccles, Behram Kursunoglu

THE PHYSIOLOGY AND PHYSICS OF THE FREE WILL PROBLEM

J. C. Eccles

Laboratory of Neurobiology

State University of New York at Buffalo

Amherst, New York 14226

THE THREE WORLD HYPOTHESIS OF POPPER

I will begin by brief reference to the philosophical basis of my discussion. I refer to the trialist philosophy that has been developed by Sir Karl Popper (1972) in his recent book "Objective Knowledge." As illustrated in Fig. 1, everything in existence and in experience is subsumed in one or other of three worlds: World 1, the world of physical objects and states; World 2, the world of states of consciousness and subjective knowledge of all kinds; World 3, the world of man-made culture, comprising the whole of objective knowledge. Furthermore it is postulated that there is interaction between these worlds. There is reciprocal interaction between Worlds 1 and 2, and between Worlds 2 and 3 via the mediation of World 1. The objective knowledge of

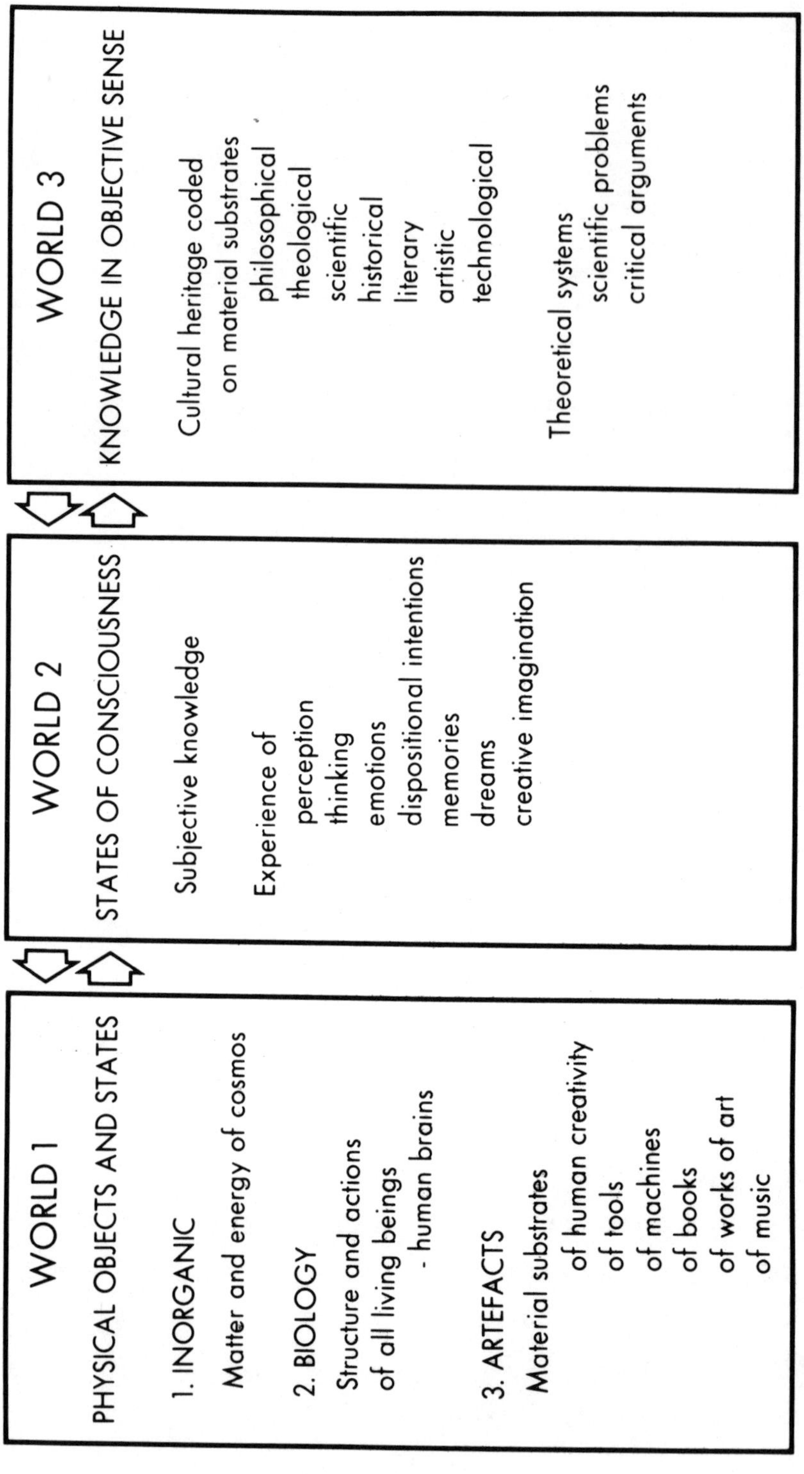

WORLD 1
PHYSICAL OBJECTS AND STATES

1. INORGANIC
Matter and energy of cosmos

2. BIOLOGY
Structure and actions
of all living beings
- human brains

3. ARTEFACTS
Material substrates
of human creativity
of tools
of machines
of books
of works of art
of music

WORLD 2
STATES OF CONSCIOUSNESS

Subjective knowledge

Experience of
perception
thinking
emotions
dispositional intentions
memories
dreams
creative imagination

WORLD 3
KNOWLEDGE IN OBJECTIVE SENSE

Cultural heritage coded
on material substrates
philosophical
theological
scientific
historical
literary
artistic
technological

Theoretical systems
scientific problems
critical arguments

Figure 1 World of consciousness. The three postulated components in the world of consciousness together with a tabulated list of their components.

World 3 (the man-made world of culture) is encoded on various objects of World 1 -- books, pictures, structures, machines -- and can be perceived only when projected to the brain by the appropriate receptor organs and afferent pathways. Reciprocally the World 2 of conscious experience can bring about changes in World 1, in the first place in the brain, then in muscular contractions, by World 2 acting on World 1. This is the postulated operation of free will that is the subject of this lecture. We may diagram the postulated interactions of this trialist interactionist hypothesis as: World 1 $\rightleftarrows$ World 2 and World 3 $\rightleftarrows$ World 1 $\rightleftarrows$ World 2, where World 2 $\rightarrow$ World 1 contains the free will problem.

Fig. 2 defines the free will problem more succinctly in terms of the three major components that are generally recognized for World 2 (cf. Polten, 1973). There is firstly the outer sense which relates specifically to the perceptions given immediately by the inputs of the sense organs, visual, auditory, tactile, smell, taste, pain, etc. Secondly, there is inner sense which comprises a wide variety of cognitive experiences: thoughts, memories, intentions, imaginings, emotions, feelings, dreams. Thirdly, and at the core of World 2, there is the self or pure ego that is the basis of the personal continuity that each of us experiences throughout our life time, spanning for example the diurnal gaps of consciousness in sleep. Each day consciousness

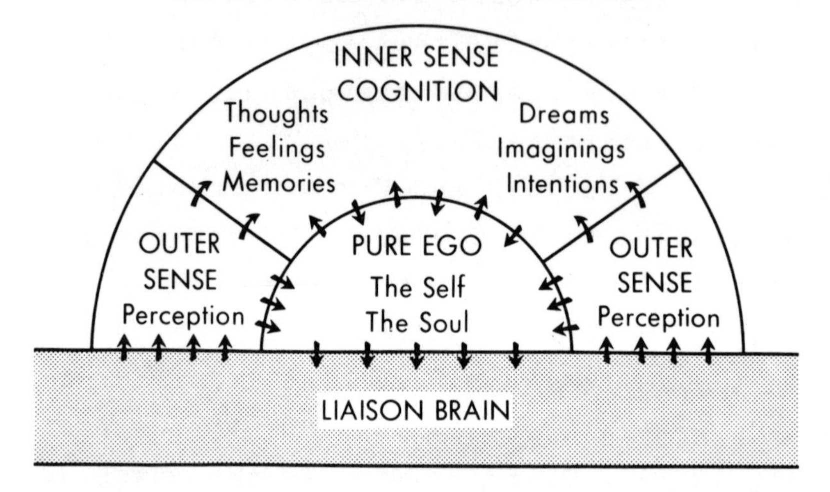

Fig. 2. Diagrammatic representation of brain-mind interaction. The three components of World 2, outer sense, inner sense and pure ego are shown with arrows depicting interactions. Also shown are the postulated interactions with the liason brain.

returns to us with its continuity essentially unbroken by the hours of unconsciousness in sleep.

FREE WILL VERSUS DETERMINISM

Sherrington (1940) in his Gifford Lectures ("Man on His Nature") has written most movingly on the self:

> "This 'I', this self, which can so vividly propose to 'do', what attributes as regards 'doing' does it appear to itself to have? It counts itself as a 'cause'. Do we not each think of our 'I' as a 'cause' within our body?"

Polten (1973) gives a good definition of an action brought about by free will.

> "An action to be free must be conscious, purposive, follow open alternative choices; and it by no means follows, as empiricist philosophers always maintain, that because it could be otherwise it need be arbitrary, or because it is not mechanically caused, it is not caused at all."

That we have free will is a fact of experience. Furthermore I state emphatically that to deny free will is neither a rational nor a logical act. This denial either presupposes free will for the deliberately chosen response in making that denial, which is a contradiction, or else it is merely the automatic response of a nervous system built by genetic coding and moulded by conditioning. One does not conduct a rational argument with a being who makes the claim that all its responses are reflexes, no matter how complex and subtle the conditioning. Nevertheless, despite these logical problems, it is widely held that free will must be rejected on logical grounds. The question can be raised: can our belief in free will be accommodated in a deterministic universe?

The distinguished physicist, Arthur Holly Compton wrote in his "Freedom of Man" (1935)

> "Is man a free agent? If... the atoms of our bodies follow physical laws as immutable as the motions of the planets, why try? What difference can it make how great the effort if our actions are already predetermined by mechanical laws...?"

As Popper (1972) comments:

> "Compton describes here what I shall call 'the nightmare of the physical determinist.'... Thus all our thoughts, feelings, and efforts can have no practical influence upon what happens in the physical world: they are, if not mere illusions, at best superfluous byproducts ('epiphenomena') of physical events."
>
> "Physical determinism, we might say in retrospect, was a daydream of omniscience which seemed to become more real with every advance in physics until it became an apparently inescapable nightmare."
>
> "...according to determinism, any theories--such as, say, determinism--are held because of a certain physical structure of the holder (perhaps of his brain). Accordingly we are deceiving

> ourselves (and are physically so determined as to deceive ourselves) whenever we believe that there are such things as arguments or reasons which make us accept determinism. Or in other words, physical determinism is a theory which, if it is true, is not arguable, since it must explain all our reactions, including what appear to us as beliefs based on arguments, as due to purely physical conditions. Purely physical conditions, including our physical environment, make us say or accept whatever we say or accept."

When discussing causality Max Planck (1936) made a statement that is relevant to this initial discussion.

> "The question of free will is one for the individual consciousness to answer: it can be determined only by the ego. The notion of free will can mean only that the individual feels himself to be free, and whether he does so in fact can be known only to himself."

There seemed to be the chance of some relief from this logical confrontation when Heisenberg made his fundamental contribution known as the Principle of Uncertainty, according to which there is a correlated uncertainty in measurements of the position and of the momentum of a particle. The more accurate is one measurement the less accurate is the other. I shall later refer to the tentative proposals made by Eddington in his attempts to utilize the uncertainty in order to resolve the problem of how free will can be accommodated to the world of physics.

THE NEUROPHYSIOLOGY OF FREELY WILLED ACTIONS

My position is that I have the indubitable experience that by thinking and willing I can control my actions if I so wish, although in normal waking life this prerogative is exercised but seldom. I am not able to give a scientific account of how thought can lead to action, but this failure serves to emphasize the fact that our present physics and physiology are too primitive for this most challenging task of resolving the antinomy between our experiences and the present primitive level of our understanding of brain function. When thought leads to action, I am constrained, as a neuroscientist, to postulate that in some way, completely beyond my understanding, my thinking changes the operative patterns of neuronal activities in my brain. Thinking thus comes to control the discharges of impulses from the pyramidal cells of my motor cortex and so eventually the contractions of my muscles and the behavioural patterns stemming therefrom. A fundamental neurological problem is: how can willing of a muscular movement set in train neural events that lead to the discharge of pyramidal cells of the motor cortex and so to activation of the neural pathway that leads to the muscle contraction?

We are now in a position to consider the experiments of Kornhuber and associates (Deecke,

Scheid and Kornhuber, 1969; Kornhuber, 1973) on the electrical potential generated in the cerebral cortex prior to the carrying out of a willed action. There is an attractive parallelism between these beautifully simple experiments and those of Galileo in investigating the laws of motion of the universe by studying the movements of metal balls rolling down an inclined plane.

The problem is to have an elementally simple movement executed by the subject entirely on his own volition, and yet to have accurate timing in order to average the very small potentials recorded from the surface of the skull. This has been solved by Kornhuber and his associates who use the onset of muscle action potentials involved in the movement to trigger a reverse computation of the potentials up to 2 sec before the onset of the movement. The movement illustrated was a rapid flexion of the right index finger. The subject initiates these movements "at will" at irregular intervals of many seconds. In this way it was possible to average 250 records of the potentials evoked at each of several sites over the surface of the skull, as shown in Fig. 3 for the three upper traces. The slowly rising negative potential, called the _readiness potential_, was observed as a negative wave with unipolar recording over a wide area of the cerebral surface (recorded by scalp leads), but there were small positive potentials of similar time course over the most anterior and basal regions of the cerebrum.

Fig. 3. Cerebral potentials, recorded from the human scalp, preceding voluntary rapid flexion movements of the right index finger. The potentials are obtained by the method of reverse analysis. Eight experiments on different days with the same subject; about 1000 movements per experiment. Upper three rows: monopolar recording, with both ears as reference; the lowermost trace is a bipolar record, left versus right precentral hand area. The readiness potential starts about 0.8 sec prior to onset of movement; it is bilateral and widespread over precentral (L. prec, R. prec) and parietal (Mid-par) areas. The premotion positivity, bilateral and widespread too, starts about 90 msec before onset of movement. The motor potential appears only in the bipolar record (L/R prec), it is unilateral over the left precentral hand area, starting 50 msec prior to onset of movement in the electromyogram. (Kornhuber, 1973).

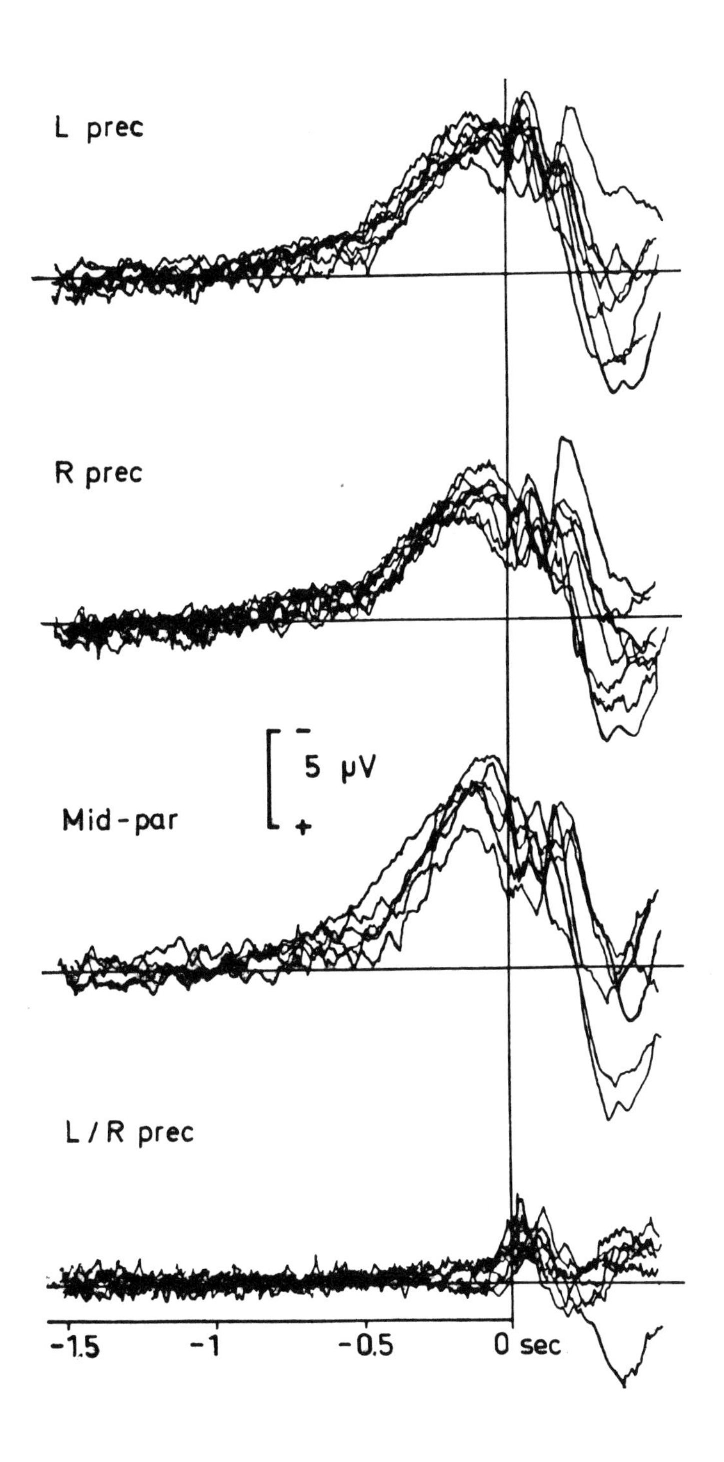
L prec
R prec
5 μV
Mid-par
L / R prec
-1.5
-1
-0.5
0 sec

Usually the readiness potential began almost as long as 800 msec before the onset of the muscle action potentials, and led on to sharper potentials, positive then negative, beginning at about 90 msec. In the lowest trace there was bipolar leading from symmetrical zones over the motor cortex, that on the left being over the area concerned in the finger movement. There was no detectable asymmetry until a sharp negativity developed at 50 msec before the onset of the muscle action potentials. We can assume that the readiness potential is generated by complex patterns of neuronal discharges that are originally symmetrical, but eventually project to the appropriate pyramidal cells of the motor cortex and synaptically excite them to discharge, so generating this localized negative wave just preceding the movement.

These experiments at least provide a partial answer to the question: What is happening in my brain at a time when a willed action is in process of being carried out? It can be presumed that during the readiness potential there is a developing specificity of the patterned impulse discharges in neurones so that eventually there are activated the pyramidal cells in the correct motor cortical areas for bringing about the required movement. The readiness potential can be regarded as the neuronal counterpart of the voluntary command. The surprising feature of the readiness potential is in its wide extent and gradual build up. Apparently, at the stage of willing a

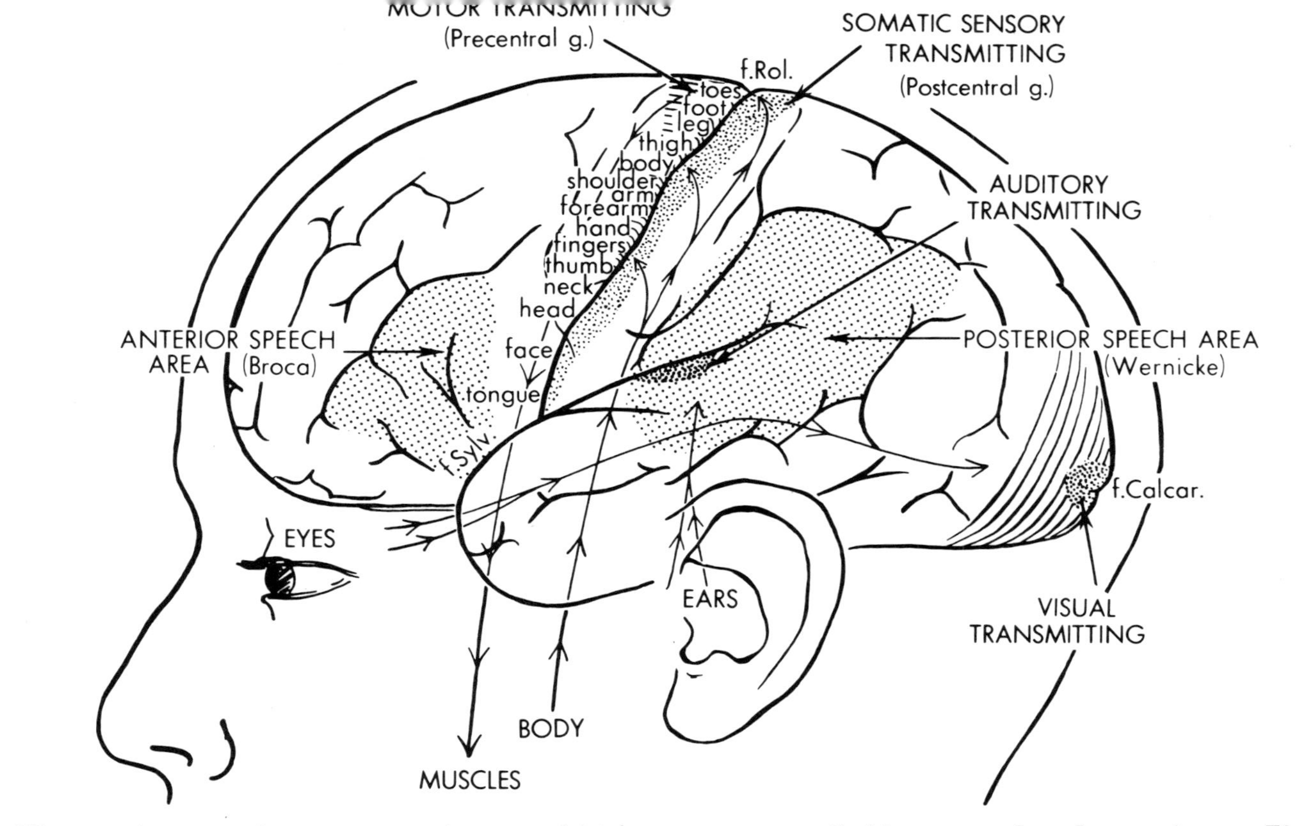

Fig. 4. The motor and sensory transmitting areas of the cerebral cortex. The approximate map of the motor transmitting areas is shown in the precentral gyrus, while the somatic sensory receiving areas are in a similar map in the postcentral gyrus. Other primary sensory areas shown are the visual and auditory, but they are largely in areas screened from this lateral view. Also shown are the speech areas of Broca and Wernicke.

movement, there is very wide influence on the patterns of neuronal operation, or as we will consider below, on the patterns of module operation. Eventually this immense neuronal activity is moulded and directed so that it concentrates on to the pyramidal cells in the proper zones of the motor cortex (Fig. 4) for carrying out the required movement. I will later continue with the neurological problems arising from these remarkable experiments.

THE UNIQUE AREAS OF THE CEREBRAL CORTEX

The evolution of man's brain from primitive hominids was associated with an amazingly rapid increase in size, from 550g to 1400g in one to two million years. But much more important was the creation of special areas associated with speech (Fig. 4). We can well imagine the great evolutionary success attending not only the growth of intelligence that accompanied brain size in some exponential relationship, but also the development of language for communication and discussion. In this manner primitive man doubtless achieved great successes in communal hunting and food gathering, and in adapting to the exigencies of life in linguistically planned operations of the community. We now know that special areas of the neocortex were developed for this emerging linguistic performance, which in 98% are in the left cerebral hemisphere (Penfield and Roberts,

1959). Usually (in 80% of brains) there is a considerable enlargement of the planum temporale in the left temporal lobe and in the areas bordering the sulcus in the inferior frontal convolution (Geschwind, 1972; Wada, 1973); and this enlargement is developed as early as the 29th week of intra-uterine life in preparation for usage some months after birth. Its development represents a very important and unique construction by the genetic instructions provided for building the human brain.

Sperry's (1968) investigations on commissurotomy patients (Fig. 5) have shown that the dominant linguistic hemisphere is uniquely concerned in giving conscious experiences to the subject and in mediating his willed actions. It is not denied that some other consciousness may be associated with the intelligent and learned behavior of the minor hemisphere. However the absence of linguistic or symbolic communication at an adequate level prevents this from being exhibited; and moreover it is unknown to the speaking subject who is in liaison with the dominant hemisphere. There is no recognizable "self" in relation to whatever consciousness there may be in liaison with the minor hemisphere. It is not therefore "self-consciousness". The situation is equivalent to the problem of animal consciousness, to which we should be agnostic.

Fig. 6 shows in diagrammatic form the association of linguistic and ideational areas of the

Fig. 5. Schema showing the way in which the left and right visual fields are projected onto the right and left visual cortices, respectively, due to the partial decussation in the optic chiasma. The schema also shows other sensory inputs from right limbs to the left hemisphere and that from left limbs to the right hemisphere. Similarly, hearing is largely crossed in its input, but olfaction is ipsilateral. (From Sperry, 1968).

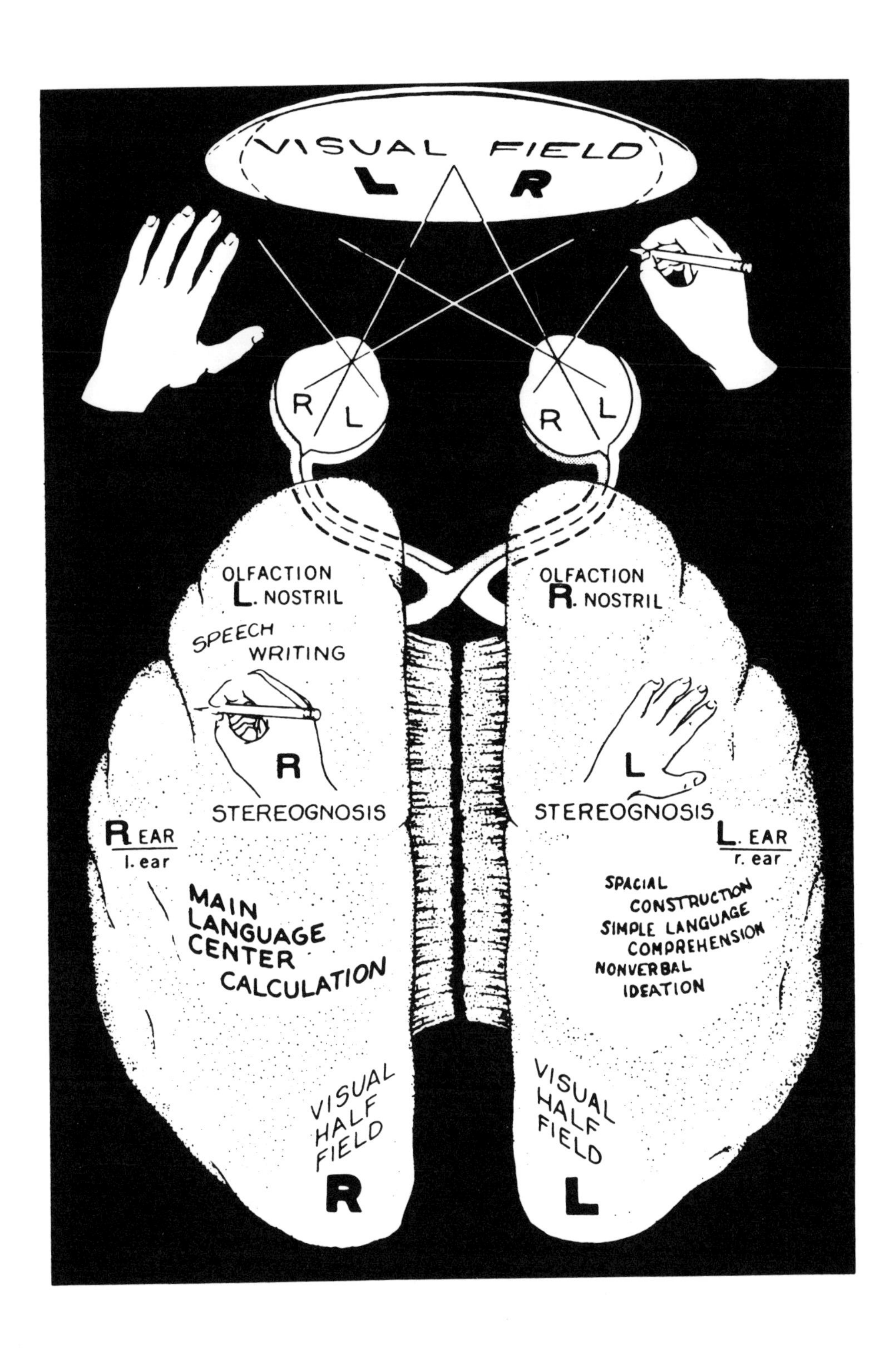
VISUAL FIELD
L
R
R
L
R
L
OLFACTION
L. NOSTRIL
SPEECH
WRITING
R
STEREOGNOSIS
R EAR
l. ear
MAIN
LANGUAGE
CENTER
CALCULATION
VISUAL
HALF
FIELD
R
OLFACTION
R. NOSTRIL
L
STEREOGNOSIS
L. EAR
r. ear
SPACIAL
CONSTRUCTION
SIMPLE LANGUAGE
COMPREHENSION
NONVERBAL
IDEATION
VISUAL
HALF
FIELD
L

Fig. 6. Modes of interaction between hemispheres. Communications to and from the brain and within the brain. Diagram to show the principal lines of communication from peripheral receptors to the sensory cortices and so to the cerebral hemispheres. Similarly, the diagram shows the output from the cerebral hemispheres via the motor cortex and so to muscles. Both these systems of pathways are largely crossed as illustrated, but minor uncrossed pathways are also shown. The dominant left hemisphere and minor right hemisphere are labeled, together with some of the properties of these hemispheres. The corpus callosum is shown as a powerful cross-linking of the two hemispheres and, in addition, the diagram displays the modes of interaction between Worlds 1 and 2, as described in the text, and as illustrated in Fig. 2.

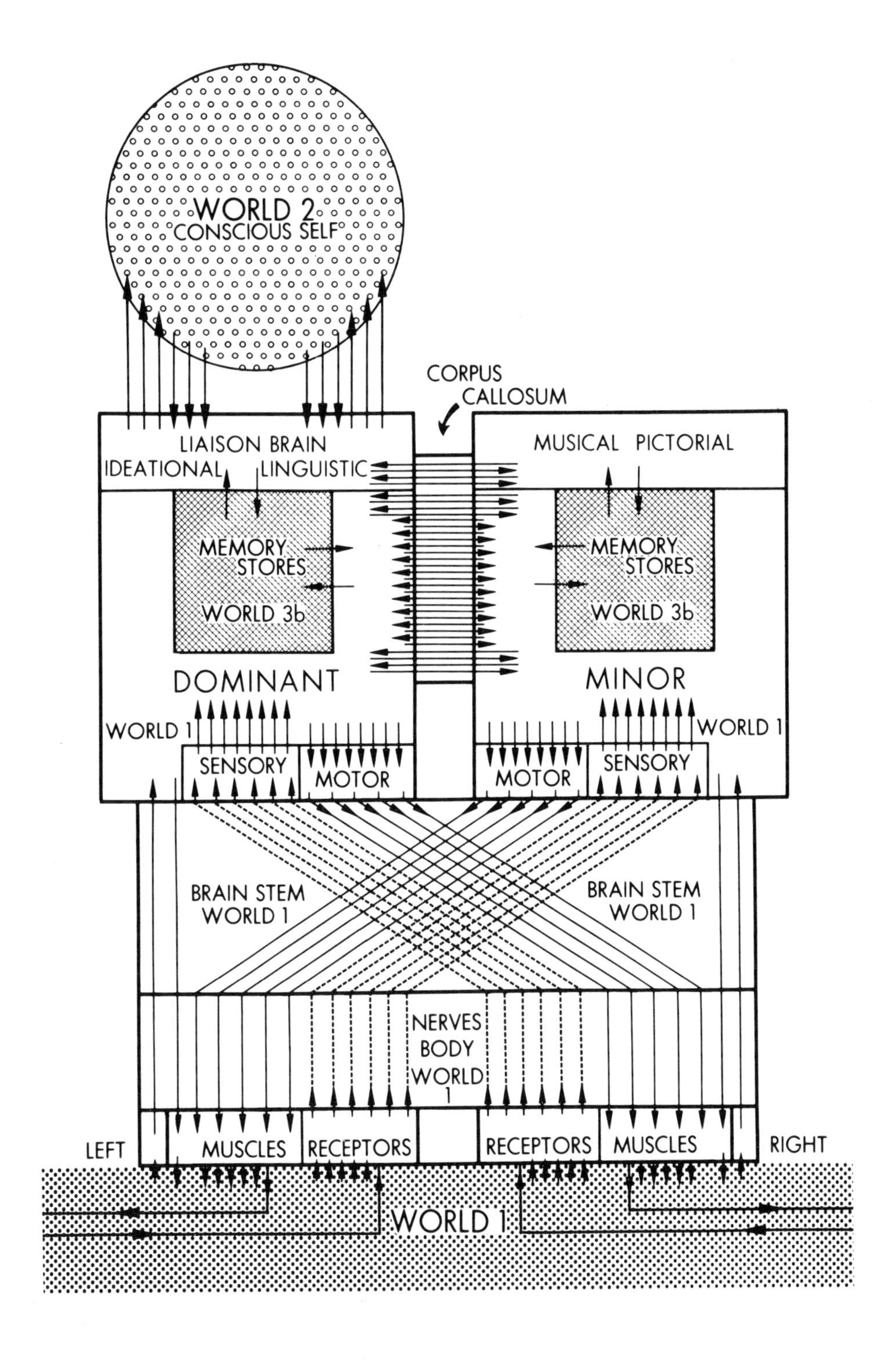
WORLD 2
CONSCIOUS SELF
CORPUS
CALLOSUM
LIAISON BRAIN
IDEATIONAL
LINGUISTIC
MUSICAL PICTORIAL
MEMORY
STORES
WORLD 3b
MEMORY
STORES
WORLD 3b
DOMINANT
MINOR
WORLD 1
WORLD 1
SENSORY
MOTOR
MOTOR
SENSORY
BRAIN STEM
WORLD 1
BRAIN STEM
WORLD 1
NERVES
BODY
WORLD
1
LEFT
MUSCLES
RECEPTORS
RECEPTORS
MUSCLES
RIGHT
WORLD 1

dominant hemisphere with the world of conscious experience. Arrows lead from the linguistic and ideational areas of the dominant hemisphere to the conscious self (World 2) (cf. Fig. 2) that is represented by the circular area above. It must be recognized that Fig. 6 is an information flow diagram and that the superior location adopted for the conscious self is for diagrammatic convenience. It is postulated that in normal subjects activities in the minor hemisphere reach consciousness only after transmission to the dominant hemisphere, which very effectively occurs via the immense impulse traffic in the corpus callosum, as is illustrated in Fig. 6 by the numerous arrows. Complementarily, as will be discussed in full later, it is postulated that the neural activities responsible for voluntary actions mediated by the pyramidal tracts normally are generated in the dominant hemisphere by some willed action of the conscious self (see downward arrows from World 2 to the liaison brain in Figs. 2 and 6). When destined for the left side, there is transmission to the minor hemisphere by the corpus callosum and so to the motor cortex of that hemisphere.

It must be recognized that this transmission via the corpus callosum is not a simple one-way transmission. The 200 million fibers must carry a fantastic wealth of impulse traffic in both directions. In the normal operation of the cerebral hemisphere, activity of any part of a hemisphere is as effectively and rapidly transmitted to the other hemisphere as to another lobe of the same hemisphere. The whole cerebrum thus achieves a most effective unity. It will be appreciated from Fig. 4 that section of the corpus callosum gives a unique and complete cleavage of this unity. The neural activities of the minor hemisphere are isolated from those cerebral areas that give and receive from the conscious self. The conscious subject is recognizably the same subject or person that existed before the brain-splitting operation and retains the unity of self-consciousness or the mental singleness that he experienced before the operation. However, this unity is at the expense of unconsciousness of all the happenings in the minor (right) hemisphere.

The diagram of Fig. 2 gives the basis for defining the postulated mode of operation of free will, which is represented symbolically by the arrows stemming from the pure ego or self. Polten (1973) states:

> "The pure ego is the necessary ingredient which changes determination to _self_-determination or libertarianism. If we did not have such an 'unmoved mover' (and it must be the _core_ of that which makes up the self!) then we could not master our environment with science and technology, as we undeniably do. Those who uphold free will need a pure ego, and the meaning and existence of free will has been so notoriously unclear and vexing largely because the meaning and existence of the pure ego has so far been

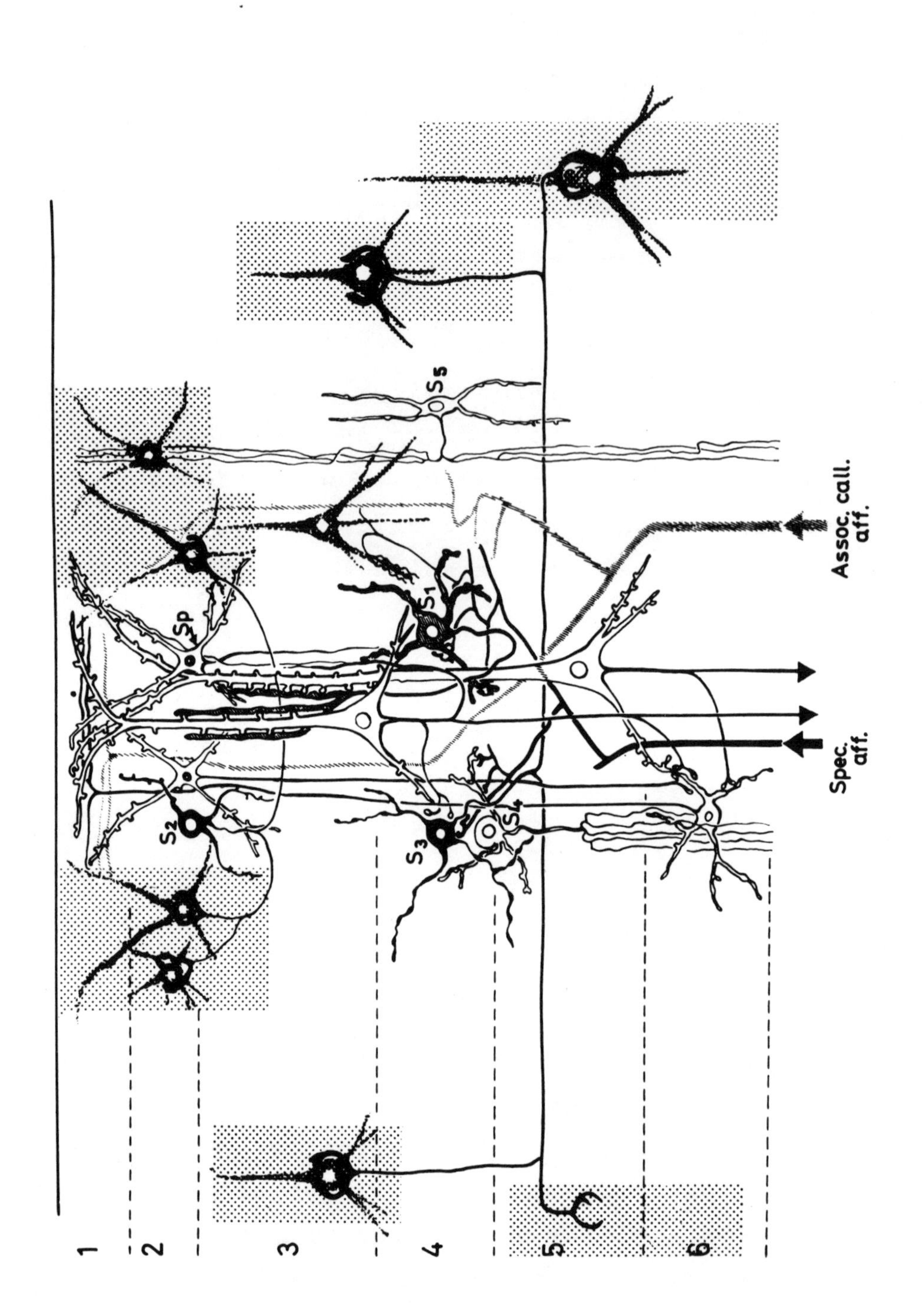
1
2
3
4
5
6
Sp
S_1
S_2
S_3
S_4
S_5
Assoc. call. aff.
Spec. aff.

Fig. 7. Semidiagrammatic drawing of some cell types of the cerebral cortex with interconnections as discussed in the text. Two pyramidal cells are seen centrally in laminae 3 and 5. The specific afferent fiber is seen to excite a stallate interneurone S_1 (cross-hatched) whose axon establishes cartridge-type synapses on the apical dendrites. The specific afferent fiber also excites a basket-type stellate interneurone S_3, that gives inhibition to pyramidal cells in adjacent columns, as indicated by shading. Another interneurone is shown in lamina 6 with ascending axon, and S_5 is an interneurone also probably concerned in vertical spread of excitation through the whole depth of the cortex. (Szentagothai, 1969).

so unclear. Thus human reflex actions, such as a knee jerk, are unfree because the pure ego is not involved; but conscious thoughts and purposive actions are free because the pure ego directs them."

STRUCTURAL AND FUNCTIONAL CONCEPTS OF THE CEREBRAL CORTEX

THE MODULAR CONCEPT

Physiological investigations by Mountcastle (1957) on the somesthetic cortex and by Hubel and Wiesel (1962) on the visual cortex revealed that the pyramidal cells of small sharply defined areas exhibited an approximately similar response to specific afferent inputs. The cells were located in cortical zones forming columns orthogonal to the cortical surface. In fact the primary sensory areas are composed of a mosaic of such columns with irregular cross sections averaging about 0.2 mm^2 in area. Recent investigations by Szentagothai (1969, 1972, 1973) have revealed that the column or module is the basic unit of the cortex and is a complex organization of many specific cell types (Fig. 7). The modules represent what he calls the basic neurone circuits that in elemental form are constituted by input channels (afferent fibers), complex neuronal interactions in the module, and output channels, largely the axons of the pyramidal cells.

In the first place the functional uniqueness

of a module (Fig. 7) derives from the limited transverse range, 500 u in laminae 3, 4 and 5, of the excitatory action by the specific and other afferent fibers, and from the powerful and vertically localized excitation by the interneurones (S_1, S_5) giving the cartridge type synapses. A further defining factor is the inhibitory surround that is built up by the basket cells (S_3) in lamina 4 acting upon pyramidal cells of adjacent modules. Also shown in Fig. 7 are the finer grain excitatory and inhibitory interactions in laminae 1 and 2. It should be noted in parenthesis that Szentagothai (1972) generalizes from the specific sensory areas to the neocortex in general.

The excitatory level built up in a module is continuously communicated to other modules by the impulse discharges along the association fibers formed by the axons of pyramidal cells and of the stellate pyramidal cells (Szentagothai, 1972). In this way powerful excitation of a module will spread widely and effectively to other modules. There is as yet no quantitative data on module operation. However the number of neurones in a module is surprisingly large--up to 10,000, of which there would be some hundreds of pyramidal cells and many hundreds of each of the other species of neurones. The operation of a module can be imagined as a complex of circuits in parallel with summation of hundreds of convergent lines onto neurones and in addition a mesh of feed-forward and feed-back excitatory and

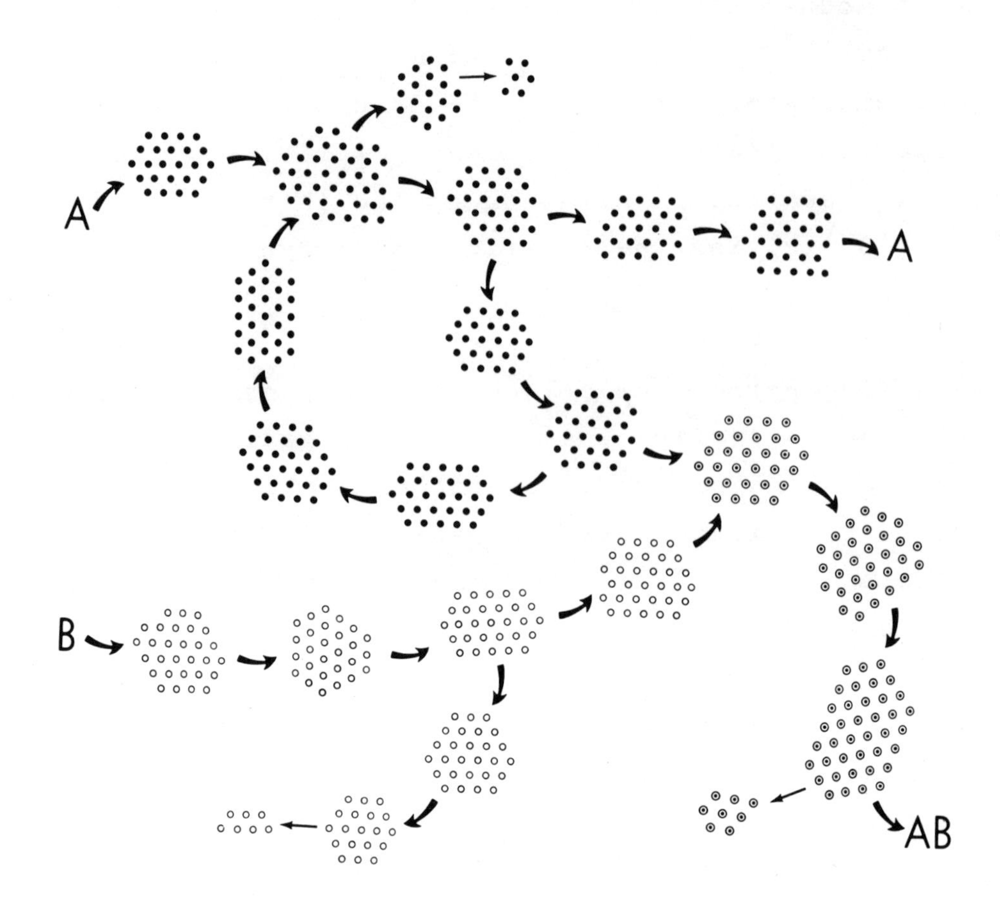

Fig. 8. In this schema of the cerebral cortex looked at from above, the large pyramidal cells are represented as circles, solid or open, that are arranged in clusters, each cluster corresponding to a column as diagrammed in Fig. 7, where only two large projecting pyramidal cells are shown of the hundreds that would be in the column. The large arrows symbolize impulse discharges along hundreds of lines in parallel, which are the mode of excitatory communication from column to column. Two inputs, A and B, and two outputs, A and AB, are shown. Further description in text.

inhibitory lines overpassing the simple neuronal circuitry expressed in Fig. 7. Thus we have to envisage levels of complexity in the operation of a module far beyond anything yet conceived, and of a totally different order from any integrated microcircuits of electronics, which would be the closest physical analogy. Moreover there will be an enormous range in the output from a module--from high frequency discharges in the hundreds of constituent pyramidal cells to the irregular low level discharges characteristic of cerebral cortex in the resting state (Evarts, 1964, 1968). The range of projection of the pyramidal cells is enormous--some go only to nearby modules, others are remote association fibers, and yet others are commissural fibers traversing the corpus callosum to areas of the other side, which tend to be in mirror-image relationship.

THE PATTERNS OF MODULE INTERACTION

Fig. 8 is a diagrammatic attempt to illustrate in the limited time span of a fraction of a second the on-going module to module transmission. It attempts to show the manner in which association fibers from the pyramidal cells in a module can activate other modules by projections of many pyramidal axons in parallel. These other modules in turn project effectively to further modules. In this assumed plan of a small zone of the neocortex the pyramidal cells of the modules are represented as circles, solid or open, according as they participate in one or another class of modality operation, e.g. to one type of sensory input for A and to another for B. Main lines of communication between successive modules are shown by arrows, and there is one example of a return circuit giving a loop for sustained operation in the manner of the closed self-reexciting chains of Lorente de No. In addition convergence of the modules for A and B modalities gives activation of modules by both A and B imputs with a corresponding symbolism--dense-core circles. The diagram is greatly simplified because in it one module at the most projects to two other modules, whereas we may suppose that in the cerebral cortex the projection is to tens or hundreds. There are 3 examples where excitation of modules was inadequate for onward propagation. Thus in the diagram two inputs A and B give only two outputs A and AB. Fig. 8 represents the kind of patterning of neuronal activation in the cerebral cortex that was imagined by Sherrington (1940). He likened it to "an enchanted loom, weaving a dissolving pattern, always a meaningful pattern, though never an abiding one, a shifting harmony of subpatterns."

The diagram of Fig. 8 is particularly inadequate in that there is no representation of the irregular background discharge of all types of cortical neurones. The modular activation and transmission must be imagined as being superimposed upon this on-going background noise.

Effective neuronal activity is ensured when there is in-parallel activity of many neurones with approximately similar connections. Signals are in this way lifted out of noise. Thus, instead of the simplicity indicated in Figs. 7 and 8, we have to envisage an irregular seething activity of the whole assemblages of neurones, the signals being superimposed on this background by phases of collusive activity of neurones in parallel either within modules or between modules.

One can surmise that, from the extreme complexity and refinement of its modular organization, there must be an unimagined richness of properties in the active cerebral cortex. It is postulated that, in a situation where the pure ego is operative (cf. Fig. 2), there will be changed patterns of modular interaction leading eventually to a change in the spatio-temporal pattern of influences playing upon the pyramidal cells in the motor cortex. The "readiness potential" (Fig. 3) bears witness to this cortical activity preceding the pyramidal tract discharge.

Evidently we have here a fundamental problem that transcends our present neurophysiological concepts. Some tentative suggestions have been made (Eccles, 1953, 1970). It is necessary to take into account the evidence that the pure ego can act in some patterned manner on cortical modules only when the cerebral cortex is at a relatively high level of excitation. Originally it was suggested that the liaison between mind and brain depended on the "mind influences" being able to modify the discharge of neurones that were critically poised at firing level. In the light of the modular concept a more attractive hypothesis would be that the modules themselves are the detector units for causal input from the pure ego. We may give them a function crudely analogous to radio-receiving units. However, evidence from patients with cerebral lesions reveals that only a special zone of the cerebral cortex would have this modular detector competence. Not only is it restricted to the dominant hemisphere, but it is further restricted to the linguistic and ideational areas, as indicated in Fig. 4.

Thus, the neurophysiological hypothesis is that the causal action of the pure ego modifies the spatio-temporal activity in the modules of the liaison zone of the dominant hemisphere. It will be noted that this module detector hypothesis assumes that the pure ego has itself some spatio-temporal patterned character in order to allow it this operative effectiveness.

This hypothesis is closely related to recent postulates by Sperry (1969) who states:

> "In the present scheme the author postulates that the conscious phenomena of subjective experience do interact on the brain processes exerting an active causal influence. In this view consciousness is conceived to have a directive role in determining the flow pattern of cerebral excitation."

"Conscious phenomena in this scheme are conceived to interact with and to largely govern the physiochemical and physiological aspects of the brain process. It obviously works the other way round as well, and thus a mutual interaction is conceived between the physiological and the mental properties. Even so, the present interpretation would tend to restore mind to its old prestigious position over matter, in the sense that the mental phenomena are seen to transcend the phenomena of physiology and biochemistry."

Just because World 2 is drawn located above the brain in Figs. 2 and 6, I do not wish to imply that World 2 is floating above the brain and has an autonomous existence and performance independent of the liaison area of the brain! On the contrary it is, so far as we can discover, intimately linked with neuronal activity in the liaison area. If that stops, unconsciousness supervenes. As shown by the arrows in both directions in Figs. 2 and 6, there is an incessant interplay in the interaction between World 2 and the liaison brain, but we know nothing about the nature of this interaction.

THE PHYSICAL IMPLICATIONS OF THE FREE WILL HYPOTHESIS

When considering the manner in which mind could operate on matter, Eddington (1939) discussed two hypotheses.

(i) It was postulated that mind could control the behavior of matter within the limits imposed by Heisenberg's Principle of Uncertainty (cf. Eddington, 1935). If q is the coordinate of a particle and p is the corresponding momentum, (the uncertainty of q) x (the uncertainty of p) = h (the quantum constant). Eddington rejected this partly because the permitted range of uncertainty would be exceedingly small. Presumably he was thinking of an object as large as a neurone. However, a neurophysiologist would now consider the much smaller synaptic vesicle as the key structure on which a "mind influence" might work. The synaptic vesicle is approximately a sphere 400 Å in diameter and so would have a mass of about 3×10^{-17} g. If, as Eddington implies, the uncertainty obtains for an object of this size, then it may be calculated that there is an uncertainty in the position of such an object of about 50 Å in 1 millisecond. These values are of interest since 50 Å is approximately the thickness of the presynaptic membrane across which the vesicle discharges its content of specific transmitter substance. However there would be a much smaller uncertainty for the particle whose movements were impeded in a viscous medium (D.L. Wilson, personal communication). Eddington (1939) later rejected this hypothesis for the further reason that it involved a fundamental inconsistency. First, behavior according to chance was postulated in making a calculation of the permitted limits according to the uncertainty principle, then it was restricted or controlled by a non-chance or volitional action (the mind influence), which necessarily must be introduced if mind is to be able to take advantage of the latitude allowed by the uncertainty.

(ii) As a consequence of this rejection, Eddington was led to an alternative hypothesis of a correlated behavior of the individual particles of matter, which he assumed to occur for matter in those special areas of the brain that are in liaison with mind. The behavior of such matter would stand in sharp contrast to the uncorrelated or random behavior of particles that is postulated in physics, and, as he stated, may be "regarded by us as something 'outside physics'" (Eddington, 1939).

The module detector hypothesis of mind-brain liaison has the merits of relating the occasions when the mind can operate on the brain to the high level of neuronal and modular activity during consciousness, and of showing how an effective action could be secured by a spatio-temporal pattern of minute "influences." If the neuronal activity of the cerebral cortex is at too low a level, then liaison between mind and brain ceases. The subject is unconscious as in sleep, anesthesia, coma. Perception and willed action are no longer possible. Furthermore, if a large part of the cerebral cortex is in the state of the rigorous driven activity of a convulsive seizure, there is a similar failure of brain-mind liaison, which is likewise explicable by the deficiency of the sensitive detectors, the critically poised neuronal-module complex.

In conclusion, I accept Popper's statement (1972) in rejecting the first hypothesis proposed by Eddington.

> "If determinism is true, then the whole world is a perfectly running flawless clock, including all clouds, all organisms, all animals, and all men. If, on the other hand, Peirce's or Heisenberg's or some other form of indeterminism is true, then sheer _chance_ plays a major role in our physical world. _But is chance really more satisfactory than determinism_?"

Popper goes on to say that the doctrine of chance

> "seems to hold good for the quantum-theoretical models which have been designed to explain, or at least to illustrate, the possibility of human freedom. This seems to be the reason why these models are so very unsatisfactory."
>
> "Compton himself designed such a model, though he did not particularly like it. It uses quantum indeterminacy, and the unpredictability of a quantum jump, as a model of a human decision of great moment."

Popper further comments that

> "Compton's postulate of freedom restricts the acceptable solutions of our two problems by demanding that they should conform to _the idea_

of combining freedom and control, and also to the idea of a 'plastic control', as I shall call it in contradistinction to a 'cast-iron control'."

A related statement was made by Born (1949)

"Since ancient times philosophers have been worried how free will can be reconciled with causality, and after the tremendous success of Newton's deterministic theory of nature, this problem seemed to be still more acute. Therefore, the advent of indeterministic quantum theory was welcomed as opening a possibility for the autonomy of the mind without a clash with the laws of nature."

"...you have now at your disposal the modern indeterministic philosophy of nature. You can assume a certain 'freedom', i.e. deviation from the deterministic laws, because these are only apparent and refer to averages. Yet if you believe in perfect freedom you will get into difficulties again, because you cannot neglect the laws of statistics which are laws of nature."

CONCLUSIONS

It has not been possible to formulate a precise hypothesis of the manner in which "mind influences" could act on the brain effectively to change the patterns of neuronal activity and hence to bring about a willed action; nevertheless there have been important advances from the classical mind-body problem, which even to this day goes by that archaic name in philosophical discussions. Ever since Descartes the problem has been known to scientists as the mind-brain problem. Now the specification can be much sharpened--to mind $\rightleftarrows$ cerebral hemisphere--to mind $\rightleftarrows$ dominant cerebral hemisphere--to mind $\rightleftarrows$ liaison areas of the dominant cerebral hemisphere. These liaison areas are especially related to the linguistic areas in the widest sense and also to adjacent areas of the dominant hemisphere that are referred to as the ideational areas. It is further recognized that mind-brain liaison can occur only when the liaison areas are in the correct level of activity, failing when this is too low or too high.

Neuroanatomical understanding of the cerebral cortex now leads to the modular concept, each module having an integrational center with an extremely complex functional design. It is postulated that in the evolution of the linguistic areas of the human brain, there was emergence of unique and even transcendent design in the modules of the speech and related ideational areas in order to provide the neuronal machinery for the handling of the necessary decoding and encoding that is involved in language comprehension and production. A foremost scientific challenge is to gain an understanding of these unique features of design and operation of the modules in the special areas of the human brain. But, as yet there are only tentative steps in understanding module design and operation in the mammalian brain--cat or monkey. We need to become much more sophisticated at this level before the linguistic areas of the human brain could be investigated with sufficient insight and imagination. By that time the physiology of the free will problem will have advanced to a level that may enable some understanding of the special detector properties of liaison brain. The physics of the brain-mind problem will certainly become by then a crucial issue.

REFERENCES

Born, M. Natural Philosophy of Cause and Chance. Oxford: Clarendon Press, 215 pp (1949).

Compton, A.H. The Freedom of Man. New Haven: Yale University Press (1935).

Deecke, L., Scheid, P. and Kornhuber, H.H. Distribution of readiness potential, pre-motion positivity, and motor potential of the human cerebral cortex preceding voluntary finger movements. Exp. Brain Res. 7, 158-168 (1969).

Eccles, J.C. The Neurophysiological Basis of Mind: The Principles of Neurophysiology. Oxford: Clarendon Press (1953). 314 pp.

Facing Reality. Springer-Verlag New York, Heidelberg, Berlin, 210 pp (1970).

Eddington, A.S. New Pathways in Science. Cambridge University Press, London, 230 pp (1935).

The Philosophy of Physical Science. Cambridge University Press, London (1939). 230 pp.

Evarts, E.V. Temporal patterns of discharge of pyramidal tract neurons during sleep and waking in the monkey. J. Neurophysiol. 27, 152-171 (1964).

Relation of pyramidal tract activity to force exerted during voluntary movement. J. Neurophysiol. 31, 14-27 (1968).

Geschwind, N. Language and the brain. Sci. Am. 226, 76-83 (1972).

Hubel, D.H. and Wiesel, T.N. Receptive fields, binocular interaction and functional architecture in the cat's visual cortex. J. Physiol. 160, 106-154 (1962).

Kornhuber, H.H. Cerebral cortex, cerebellum and basal ganglia: An introduction to their motor functions. In: The Neurosciences: Third Study Program. Ed. by F.O. Schmitt. New York: The Rockefeller University Press (1973).

Mountcastle, V. Modality and topographic properties of single neurones of cat's somatic sensory cortex. J. Neurophysiol. 20, 408-434 (1957).

Penfield, W. and Roberts, L. Speech and Brain-Mechanisms. Princeton, New Jersey, Princeton University Press (1959).

Planck, M. The Philosophy of Physics. London: George Allen & Unwin Ltd. 118 pp (1936).

Polten, E.P. A Critique of the Psycho-Physical Identity Theory. Mouton Publishers, The Hague (1973). 290 pp.

Popper, K.R. Objective Knowledge: An Evolutionary Approach. Oxford: Clarendon Press (1972). 380 pp.

Sherrington, C.S. Man on His Nature. London: Cambridge University Press, 413 pp (1940).

Sperry, R.W. A modified concept of consciousness. Psychol. Rev. 76, 532-536 (1969).

Szentagothai, J. Architecture of the cerebral cortex. In: Basic Mechanisms of the Epilepsies. Ed. by H.H. Jasper, A.A. Ward and A. Pope. Little, Brown & Co., Boston (1969). 13-28.
The basic neuron circuit of the neocortex.

In: Symposium on Synchronization Mechanisms. Eds. Petsche-Brasier. Springer, Vienna (1972).
Synaptology of the visual cortex. In: Handbook of Sensory Physiology. Ed. R. Jung. Vol. 7/3, Springer-Verlag (1973).

Wada, J.A., Clarke, R.J. and Hamm, A.E. Morphological asymmetry of temporal and frontal speech zones in human cerebral hemispheres: observation on 100 adult and 100 infant brains. Xth Int. Cong. Neurol., Barcelona (1973).

IS THE BRAIN A CHEMICAL COMPUTER?

G. Ungar

Baylor College of Medicine

Houston, Texas

A widely prevailing view of the nervous system today is that it operates by constant monitoring of the environment and by "simulating", in the manner of a computer, the relations of the organism with its universe. Like the previous models proposed for the nervous system (hydraulic machine, clockwork, telephone switchboard), the computer reflects the technological stage reached by society. It represents a considerable advance over its immediate predecessor, the switchboard, which could only transmit and direct the flow of information without actually "processing" it.

One can agree that the brain behaves like a "data processing machine" as computers are often called: it submits the data supplied by the input to the various operations that are programmed into it and puts out the results of these operations in the form of effector reponses. Like all computers, it can also store data and retrieve them when required. There are, however, a considerable number of differences between the brain and present-day computers. I am going to review these briefly but will devote most of this paper to one essential difference: the nervous system, instead of using electronic means for its computations, operates on chemical principles.

BRAIN VS. COMPUTER

There have been many comparisons between computer operations and biological information processing since J. von Neumann's classical essay (1958) and I shall summarize here only the principal points. Some of the characteristics of the brain are duplicated in some computers but there is no man-made machine at present that even approximates all the known features of the central nervous system.

Perhaps the most important difference is the parallel functioning of the brain as opposed to the sequential or serial operations of the computers. The brain receives simultaneous inputs from up to 10^6 sensory receptors at a given time while even the most sophisticated machines can handle only one at a time. One could almost say that a brain is an aggregate of a large number of computers connected together so that their simultaneous inputs can be analyzed, compared and correlated. This would make the brain ideally suited to be a "correlation catcher" (Fuller and Putnam, 1966).

Although this characteristic is to a certain extent compensated by the speed of the computer (a million times faster than that of the brain), it cannot provide the multiplicity of connections that allow the basic brain operation called integration, to be discussed later.

The space requirements of a computer were estimated by von Neumann to be 10^8 to 10^9 larger than those of the brain. In spite of the advances accomplished since his estimation, a computer including the same components as the brain would weigh over 10,000 tons. Energy requirements would have to be multiplied by the same factor.

An important feature of the central nervous system is its hierarchic structure. As a direct consequence of its evolutionary origin, the brain is built up of a series of levels of different phylogenetic origin, each controlling those below and being controlled by those above. This implies a considerable degree of redundancy and explains the remarkable tolerance of the brain to losses of material. This hierarchy of structure involves also a functional hierarchy and is responsible for certain peculiarities of the brain that will be discussed below.

The basic operations of the brain are genetically programmed and provide stereotyped responses to some environmental stimuli. In all higher animals, however, the brain is constantly reprogramming itself to modify its output on the basis of "learned" information. Besides the genetic program, each brain has its individual "history" that becomes an integral part of its information store.

The nervous system is not a piece of inert machinery; it has a baseline spontaneous activity that is modified by inputs. In higher organisms this spontaneous activity can be an internal source of input just as effective as the environmental stimuli.

The last point is that there is no room in biological systems for the electronic principle that animates the man-made computer. The electronic principle, of course, is by no means inherent in data processing. The first calculators were mechanical and we often hear of acoustical or optical computing devices but electronic computers are certainly the best adapted to our present technological level. I shall try to show now that computing has been going on by chemical means since life started on our globe.

CHEMICAL COMPUTATION IN BIOLOGICAL SYSTEMS

The advances of molecular biology have provided us with an outline of the process by which genetic information controls the development and functioning of the cell. The information, stored in the genome in terms of nucleotide sequences in the DNA molecule, is transcribed into equivalent sequences of RNA which, in turn, are translated into the amino acid code to build the structural and catalytic proteins of the cell. The latter activate the chemical reactions necessary to form the other cell constituents. Each of these steps includes input of information that is processed and put out in a different form. Without unduly stretching the point, one can regard enzyme activity itself as a form of computation, in which the substrate is the input and the product the output.

In the cells of the multicellular organisms there is also a more complex form of computation. The genome contains the information necessary to form all the different types of cells of which the organism is composed. Each cell, therefore, uses only a fraction (about 10 to 20%) of its genes, so that the rest of the genome is inhibited, "repressed" by a combination of the active groups of DNA with a basic protein, a histone. To activate or "derepress" the genes relevant to the given cell type there is a complex computing device that is still only partially understood. A similar mechanism can, under the appropriate stimulus, "induce" additional genes necessary for survival. These mechanisms assure that a liver cell will not grow into a neuron or vice-versa.

These examples of chemical computation should suffice to illustrate the way in which biological systems handle information, by converting it into molecular structures which can then be processed by the chemical machinery of living matter.

It is now widely recognized that living

systems are essentially chemical machines; they derive their energy from chemical reactions and they are controlled by chemical agents capable of combining with their constituents. For many years, the nervous system seemed to be exempted from this generalization. Largely for historical reasons, it was regarded as a physical, more specifically, an electrical machine. Its metabolic needs were disregarded and the evidence for a chemical mediation for synaptic transmission was rejected for over twenty years.

It is now time to recognize that neurons are merely highly specialized and differentiated cells and are just as fully based on chemical principles as all the other cells. The bioelectric phenomena are common to all cells and are caused by a peculiar distribution of ions between living matter and its environment.

CHEMICAL COMPUTATION IN THE NEURON

A nerve cell is a highly polarized structure: it receives information on its cell body and its dendritic arborizations, and after processing it, puts out the result of the computation by its axon. Most neurons receive inputs from a large number of others and their output can be distributed to many other neurons.

The output is determined by the inputs but probably not in the simple linear manner indicated by Sherrington (1906). Acceptance of an input does not depend solely on the level of the presynaptic potential. At each synapse, a complex computation determines acceptance, based on the amount of transmitter released by the presynaptic ending by unit of time, the availability of receptors for the given transmitter at the postsynaptic site, the activity of the catabolic enzymes present, the previous state of the site and the possibility of modulating influences. Receptors may exist in different allosteric forms: one ready to combine with the transmitters, the other with no affinity for them. Some of the input being excitatory, some inhibitory, the resulting value is then further computed somewhere in the cell to determine whether it reaches the threshold necessary to fire the axon. Here again a number of factors come into action, the genetic makeup of the cell, the hormones and other substances acting upon it, and its previous inputs.

Acceptance of the input and firing of the outgoing action potential depend ultimately on the chemical state of the proteins, lipids and probably other substances present in the membrane. It is this chemical state that determines whether the Na^+ ions are allowed to enter the cell and initiate the electrical change. Each neuron, therefore, is a highly complex chemical computer unit capable of calculating all the factors mentioned above and thereby determine whether the axon would fire or not.

COMPUTATION IN NEURAL NETWORKS

The sequence of events just described was first analyzed by Sherrington (1906) who called it integration. His analysis, however, was done in purely electrical terms, assuming that the output represented the algebraic sum of the incoming excitatory and inhibitory impulses. The Sherringtonian concept of integration has had a well deserved influence, by pointing to the information processing function of the nervous system.

Before proceeding from the functioning of single neurons to that of neuronal sets, I should like to survey briefly the present ideas on the mechanism of formation of neural networks. There is ample anatomical and physiological evidence for a highly specific organization of the brain taking place during embryonic development. Most organisms are born with their nervous system programmed to fulfill the homeostatic regulatory functions necessary for their survival, by means of ready-made physiological and behavioral responses to a set of stimuli. The pathways for these responses have been organized without any environmental input and must, therefore, be genetically determined.

The problem of the mechanism by which the genetic blueprint controls the construction of neural pathways has been particularly studied by Weiss (1955), Sperry (1963), Jacobson (1970) and Gaze (1970). The available evidence points to the probability that the neurons destined to be part of the same pathway join up with each other by means of a chemical recognition system. According to the doctrine of "chemospecificity of the pathways", neurons provided with the same or a complementary molecular "label" establish connections and organize the pathway.

It is known, of course, that synaptic transmission is accomplished by release of a chemical transmitter at the presynaptic ending and its binding to a receptor at the postsynaptic site. This is now a basic tenet of neurobiology and neurons are divided into categories according to the transmitter they release (adrenergic, cholinergic, etc.) and the transmitter for which they possess receptors (adrenoceptive, cholinoceptive, etc.). According to Dale's "law" (1948) all the branches of a given axon release the same transmitter but it is not necessarily true that each neuron is receptive only to one transmitter; it may possess post-synaptic sites provided with receptors for several transmitters.

Although concordance between transmitter and receptor is a necessary condition of synaptic transmission, it is unlikely that the recognition system that organizes the pathways is identical with the transmitter-receptor concordance. There are not enough different transmitters known to provide for the estimated 10^7 specific pathways possessed by the human brain. There is, therefore, need for another type of recognition molecules, probably proteins, that have a sufficient potential information content to label all the pathways (see the hypothesis of Roberts and Flexner, 1966).

Innate stimulus-response and behavioral patterns correspond to prewired, genetically determined neural circuits for which I proposed the name protocircuits. In each of these protocircuits the synapses are probably labeled by a specific molecular marker that gives them permanence and serves as a signpost to direct the traffic of nerve impulses along the given circuit. There have been several suggestions (Jacobson, 1969; Roberts and Flexner, 1966) that the molecular labels of the protocircuits may be used for marking the circuits through which acquired or learned behavior is expressed (called metacircuits by Barbizet, 1968).

The mechanism by which such metacircuits are created was explained as the result of the simultaneous firing or concomitant activity of neurons (Hebb, 1949) and had gained wide acceptance. Szilard (1964) proposed that the creation of new synaptic junctions by this mechanism is accompanied by a chemical process to which he gave the name "transprinting". This is the acquisition by a neuron of a label belonging to a circuit different from the one to which it belongs. Transprinting would occur because the simultaneous firing, by increasing the permeability of the synaptic membrane, would allow the transsynaptic transport of the molecular label.

In a hypothesis, inspired by Szilard's transprinting concept, I proposed (Ungar, 1968, 1970, 1972) that the molecular labels of the protocircuits involved in the formation of a new metacircuit could combine to encode the synapses situated at some critical "crossroads." Almost by definition, this combination should be created by a non-genetically determined chemical reaction, independent from the ribosomal mRNA-directed protein synthesis. There is evidence that peptide chains can be synthesized by purely enzymic processes (Lipmann, 1971; Meister, 1973; Reichlin and Mitnick, 1973) but direct proof for this reaction taking place at synapses involved in learning is still missing.

We do not know how this combination of labels fits into the computing scheme of the synapse outlined above. It may be by controlling the transmitter-receptor system or by some other modulating device. Whatever the exact mechanism may be, it is part of the synaptic computation that accepts or rejects the input of information. Isolation of peptides from the brain of animals trained for specific behaviors gives support to the hypothesis (Ungar, 1973) but a great deal of further work will be necessary to convince the skeptics.

Another example of complex chemical computation is the process of hierarchic integration which is the hallmark of higher nervous activity.

It is the process by which information gathered by sensory stimuli gets organized into percepts (first separate visual, auditory, tactile or other images) and subsequently integrated into increasingly complex concepts corresponding to the components of the universe. These concepts can then be used to establish correlations between the elements of the universe to form interacting systems. The next step, accomplished only by the human brain, is to give names to the concepts. This conversion of eidetic into symbolic information is what Pavlov called the switch to a "second signalling system." The difference also presents a parallel with the distinction von Neumann (1956) has made between a "complete code" and a "short code" in computers. It is possible that the speech centers of the left hemisphere of the human brain represent a distinct computer operating on a "short code". In von Neumann's words, "Anything that the first machine [the rest of the brain] can do in any length of time and under the control of all possible order systems of any degree of complexity may now be done as if only 'elementary' actions - basic, uncompounded, primitive orders - were involved."

In chemical terms, this may mean that the increasing complexity and size of the molecular labels corresponding to an object could be converted into a more compact "code word," resulting in a significant increase in the agility of information processing. There is no doubt that the possession of symbolic communication is the main factor in man's mental superiority. Whether the underlying cause is the simplification of the molecular coding system still remains to be proven.

CODING IN A CHEMICAL COMPUTER

I hope to have conveyed the idea that the nervous system is an aggregate of computers, each functioning simultaneously and in close connection with each other. The individual computers that enter into its composition may be simpler than the present day industrial machines but, put together, the whole system defies all comparison with them.

The multiplicity and complexity of the system suggests that, unlike the data-processing machines, the brain does not have a unique coding system. The coding scheme, i.e., the system of signals which represents the input of information is alternatively digital and analog. I was shown a long time ago that the bioelectric pulses, into which the energy of the stimulus is transduced, are frequency modulated. The intensity of the stimulus is about the only information that the brain obtains from these pulses. The qualitative element of the information, that is, whether it is visual, auditory, olfactive, etc. is determined entirely by the channel on which the impulses travel. Any stimulation of a neuron of the optic pathway will be decoded as visual, of the acoustic nerve as auditive, etc. This has been known for over 150 years as the "law of specific energies" stated by J. Müller. The specificity, however, goes further than Müller could have thought. Each particular fiber of the optic nerve corresponds to a specific point of the retina and will project to a predetermined site in the visual cortex, and each fiber of the acoustic nerve is decoded as a sound of a specific frequency. The brain operates therefore on the principle of "labeled lines" forming a structural code the specificity of which is genetically determined.

We have seen that there is a good evidence for the structural code being based on a chemical recognition system so that it may represent ultimately a program code operating on molecular principles. If the genetic program of the brain, i.e., the formation of protocircuits is coded in chemical terms, it is probable that the metacircuits, created by acquired information, are also maintained by a chemical process. Thus, there is a good possibility that both the innate program and the continuous reprogramming, characteristic of the brain, are based on the same principles.

A final evidence for the chemical operation of the nervous system is that it can be modified by chemical means. We know that hormones and neurotransmitters control the basic drives and the affective tone that motivate our feelings and our behavior. It is also well known that these can be affected by drugs whose action is now universally interpreted as being due to a chemical reaction between drug and receptor. It would be difficult to imagine that any chemical agent, short of corrosive acids or bases, could affect the operations of an electronic computer. The drugs used up to now, acting on transmitter systems, are only capable of changing the affective components of behavior. Once the molecular mechanisms of information processing become better known, we may find means to control directly the cognitive functions of the brain. Whether such a control is desirable or not, cannot be answered by the scientific method. It will be a matter for medical, social and ethical considerations.

REFERENCES

Barbizet, J., 1968, Learning and use of knowledge. In: Neurosciences Research (S. Ehrenpreis and O. C. Solnitzky, ed.), p. 316. London: Academic Press.

Dale, H. H., 1948, Transmission of effects from the endings of nerve fibers, Nature 162: 558.

Fuller, R. W., and Putnam, P., 1966, On the origin of order in behavior, General Systems 11: 99.

Gaze, R. M., 1970, The Formation of Nerve Connections. London: Academic Press.

Hebb, D. O., 1949, The Organization of Behavior. New York: Wiley.

Jacobson, M., 1969, Development of specific neuronal connections, Science 163: 543.

Jacobson, M., 1970, Developmental Neurobiology. New York: Holt, Rinehart and Winston.

Lipmann, F., 1971, Gramicidin S and tyrocidine biosynthesis: a primitive process of sequential addition of amino acids on polyenzymes. In: Chemical Evolution and the Origin of Life (R. Buvet and C. Ponnamperuma, ed.) p. 381. Amsterdam: North-Holland Publishing Co.

Meister, A., 1973, On the enzymology of amino acid transport, Science 180: 33.

Reichlin, S., and Mitnick, M., 1973, Biosynthesis of hypothalamic hypophysiotropic factors. In: Frontiers in neuroendocrinology, 1973 (W. F. Ganong and L. Martini, ed.), p. 61. New York: Oxford University Press.

Roberts, R. B., and Flexner, L. B., 1966, A model for the development of retina-cortex connections, American Scientist 54: 174.

Sherrington, C. S., 1906, The Integrative Action of the Nervous System. New Haven: Yale University Press.

Sperry, R. W., 1963, Chemoaffinity in the orderly growth of nerve fiber patterns and connections, Proc. Nat. Acad. Sci. U.S.A. 50: 703.

Szilard, L., 1964, On memory and recall, Proc. Nat. Acad. Sci. U.S.A. 51: 1092.

Ungar, G., 1968, Molecular mechanisms in learning, Perspectives Biol. Med. 11: 217.

Ungar, G., 1970, Molecular mechanisms in information processing, Int. Rev. Neurobiol. 13: 223.

Ungar, G., 1972, Molecular organization of neural information processing. In: The Structure and Function of Nervous Tissue (G. H. Bourne, ed.), Vol. 4, p. 215. New York: Academic Press.

Ungar, G., 1973, The problem of molecular coding of neural information. A critical review, Naturwissenschaften 60: 307.

Von Neumann, J., 1958, The Computer and the Brain. New Haven: Yale University Press.

Weiss, P., 1955, Nervous system. In: Analysis of Development (B. H. Willier, P. A. Weiss, and V. Hamburger, ed.), p. 346. Philadelphia: Saunders.

EVIDENCE FOR PHASE-TRANSITION IN NERVE MEMBRANE

Ichiji Tasaki

Laboratory of Neurobiology

National Institute of Mental Health

Bethesda, Maryland

The process of excitation of the nerve membrane is characterized by a sudden, large change in both the membrane potential and the membrane resistance[1]. It is possible to induce such a change in the membrane properties by non-electric means[2]. A detailed analysis of such processes of excitation by non-electric means has led us to the notion that the membrane macromolecules are capable of undergoing phase-transitions under a variety of conditions. In the present paper we review the results of recent experiments along this line[3] and discuss their interpretation based on the consideration of stability, fluctuations and phase-transitions in open systems[4].

Giant nerve fibers (axons) taken from squid caught in the northern part of the Sea of Japan were used in these experiments. After excision from the mantle, the axon was internally perfused with a dilute (15 to 30 mM) NaF solution. Initially, the external fluid medium was a 100 mM $CaCl_2$ solution. [The osmolarity of both internal and external media was maintained by addition of glycerol, the pH of the internal medium was kept at 7.3 with sodium phosphate buffer and that of the external medium at 8.0 by adding a trace of Tris-buffer.] Two Ag-AgCl agar electrodes, one inside and the other outside the axon, were used to record the potential difference across the axon membrane.

Record A in the figure shows the effect of addition of an iso-osmotic NaCl solution to the external medium upon the membrane potential. It is seen that the electric potential of the axon interior (relative to that of the external medium) rose suddenly when the external NaCl concentration reached a certain (critical) level. An additional increase in the external NaCl concentration produced only a smooth, gradual rise in the intracellular potential. In a separate experiment using an A.C. impedance bridge, it was shown that a sudden rise in the membrane potential was accompanied by a simultaneous fall in the membrane resistance. Similar results obtained with giant axons internally perfused with a dilute CsF solution were reported previously[2].

When the external NaCl concentration was below the critical level, such a sudden change in the membrane properties could be induced simply by reducing the external $CaCl_2$ concentration

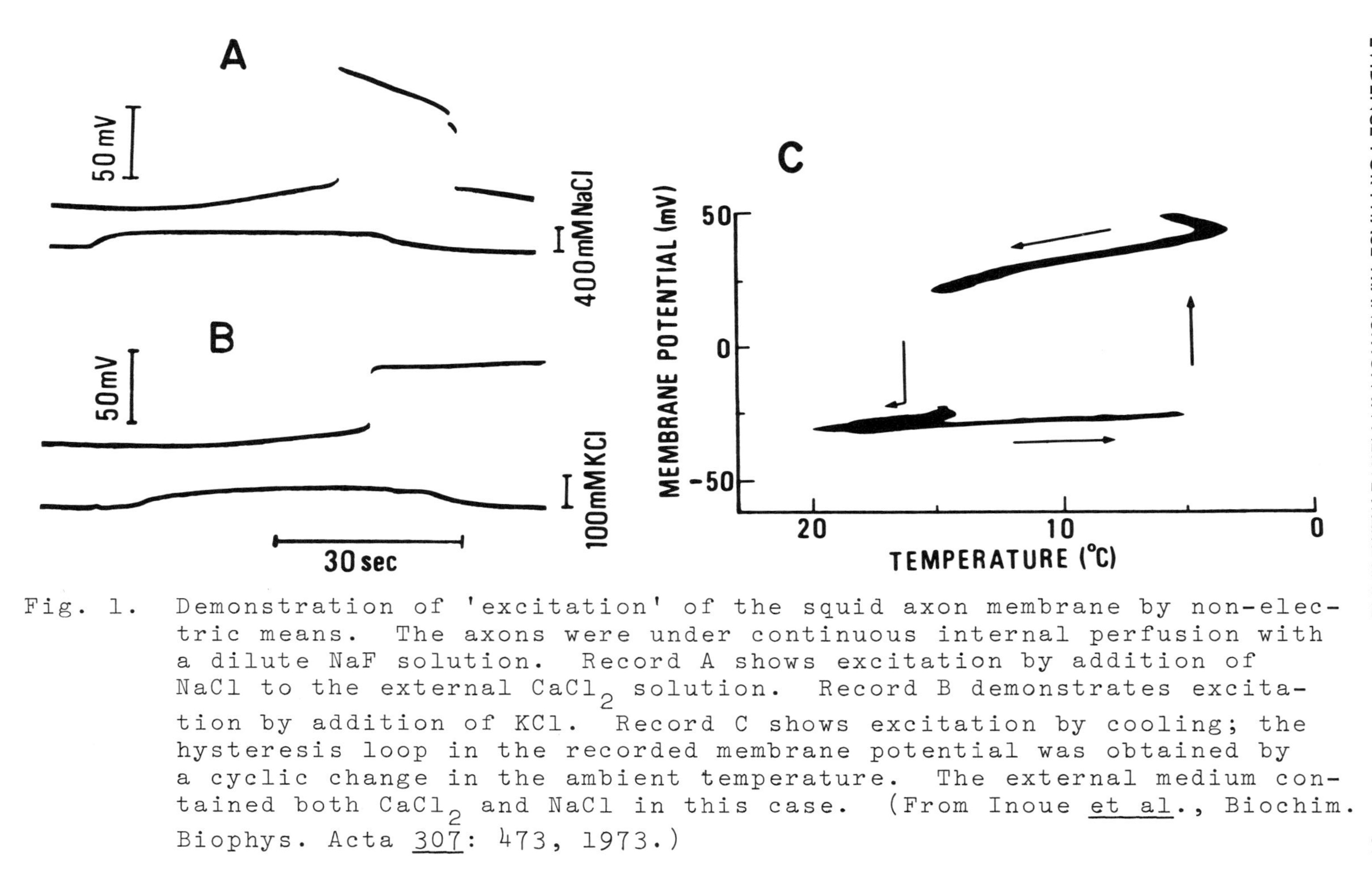

Fig. 1. Demonstration of 'excitation' of the squid axon membrane by non-electric means. The axons were under continuous internal perfusion with a dilute NaF solution. Record A shows excitation by addition of NaCl to the external $CaCl_2$ solution. Record B demonstrates excitation by addition of KCl. Record C shows excitation by cooling; the hysteresis loop in the recorded membrane potential was obtained by a cyclic change in the ambient temperature. The external medium contained both $CaCl_2$ and NaCl in this case. (From Inoue _et al_., Biochim. Biophys. Acta _307_: 473, 1973.)

without changing the NaCl concentration. This fact indicates that the univalent-divalent cation-concentration ratio (or equivalent fraction of the univalent cations), rather than the NaCl concentration itself, plays the crucial role in production of these abrupt changes in the membrane properties. It is important to note that solutions of CsCl, RbCl or KCl could be used in place of NaCl in the experiment mentioned above. Record B in the figure was obtained by adding a KCl solution to the external $CaCl_2$ solution. The abrupt change in the membrane potential was very similar to that produced by NaCl, except that the amplitude of the potential jump was smaller in this case (Record B) than in the previous case (Record A). The critical concentration of KCl was, however, far smaller than that of NaCl. In this respect CsCl and RbCl were found to be intermediate between NaCl and KCl, CsCl being closer to NaCl.

It is well known that the membrane resistance is determined by the mobility-concentration products of the ions within the membrane. Since the mobility of Ca-ions in proteins and lipids is known to be far smaller than that for univalent cations in the same environment, a sudden fall in the membrane resistance associated with the large potential change in the membrane potential (Record A) may be interpreted as indicating a cooperative exchange of the intramembrane Ca-ions for Na-ions. Such a cooperative ion-exchange process must be associated with a large change in the ion selectivity which, together with changes in the ion mobilities, determines the sign and the magnitude of the potential jump. This interpretation leads us to the concept that the nerve membrane at rest is a Ca-rich (i.e., high-resistance) state and that, at a critical univalent-divalent cation-concentration ratio in the membrane, the membrane undergoes a transition to a univalent-cation rich (i.e., low-resistance) state. In the state of the membrane rich in univalent-cations (Na-ions in this case), the membrane resistance is low because of a high mobility of the univalent cations.

Record C in the figure was obtained by changing the temperature slowly and uniformly along the axon. The medium in which the axon was immersed contained both $CaCl_2$ and NaCl, the Na-Ca ratio being slightly smaller than the critical level required for production of an abrupt potential change. The internal perfusion fluid was, as in the preceding experiments, a dilute NaF solution. It is seen in the figure that the membrane potential changed only slightly until the temperature reached about 5°C at which there was a sudden rise in the membrane potential. When the temperature was raised slowly from about 4°C, an abrupt change in the membrane potential was encountered, not at 5°C, but at about 15°C in this case. Thus, a distinct hysteresis loop was obtained when the temperature was altered in a cyclic manner.

These thermally induced transitions can be explained in the following fashion. In inanimate cation-exchangers, it is known that an exchange of Ca-ions for Na-ions is exothermic, involving an enthalpy change of 1.5 to 3 kcal/equiv. Consistent with this well-known fact, is that the critical Na-Ca ratio in the medium at 20°C is greater than that at 4°C. When the external Na-Ca ratio is close to the critical level at room temperature, therefore, cooling of the membrane is expected to bring about a transition from the Ca-rich state to the Na-rich state.

It is easy to demonstrate that the nerve membrane at rest with a salt solution on either side is stable against small perturbations. [Electric currents, temperature changes, changes in the electrolyte composition of the media can be used to perturb the membrane.] When the membrane is brought close to the critical (transition) point, the system becomes unstable and fluctuation in the membrane potential (and resistance) increases enormously. The abrupt change in the membrane potential was found to start from one of the peaks of "giant fluctuations". The thermodynamic basis for fluctuations in dissipative systems near their transition points was discussed recently by Glansdorff and Prigogine (p. 104 in Ref. 4).

The experimental results mentioned above have led to several predictions, which were later verified experimentally. Consequently, we believe that the macromolecular hypothesis of nerve excitation which assumes phase-transitions between two stable states is strengthened to a considerable degree by these experiments.

REFERENCES

1. Cole, K. S. and Curtis, H. J. Gen. Physiol., 22: 649, 1939.
2. Tasaki, I. Nerve Excitation: A Macromolecular Approach, Charles C. Thomas, Springfield, Ill., 1968.
3. Inoue, I., Kobatake, Y. and Tasaki, I. Biochim. Biophys. Acta 307: 471, 1973.
4. Glansdorff, P. and Prigogine, I. Thermodynamic Theory of Structure, Stability and Fluctuations. Wiley-Interscience, London, 1971.

LANGUAGES, HIERARCHICAL STRUCTURES, AND LOGIC

Eduardo R. Caianiello

Laboratorio di Cibernetica del CNR

Arco Felice, Napoli
and
Istituto die Fisica

University of Salerno

I think it appropriate, and thank Professor Kursunoglu for planning it, that at a gathering of this nature the science of Cybernetics be not absent. Whether you call it this name, or "General Systems Theory", "Study of Artificial Intelligence" or any other denomination that different people use in different places to denote exactly the same kind of work, the relevant fact is that a new paradigm in science is thereby implied, in which the _global_ approach, rather than the specialistic particularization, is the center of interest. Interdisciplinary by its very nature, Cybernetics (this is the word I consider as the most fitting for many reasons, not excluding my long association with Norbert Wiener) makes the study of relations among parts of very complicated systems its main objective of research. In such situations all that physics can usually do is to resort to statistics; but this is not good enough for our purpose, because all one can get from statistics is an average behaviour and a fluctuation around that average: just to quote one example, cyclic phenomena involving large systems, for instance Circadian rhythms, which are certainly not fluctuations, cannot be predicted through the statistics of equilibrium states, and the dynamical study of irreversible processes is just beginning and in any case not sufficient to cover the whole field of interest to us.

Consider systems in which phenomena that we may term life, intelligence, or just organization, are observed: we definitely see that any such system, be it a human body, a beehive, a large corporation or a military organization, is always experimentally found to possess a hierarchical organization; that is, a many-layered structure composed of levels, the elements or individuals at each level interacting with those of the same level, exerting an influence over a larger number of individuals of the level below and being grouped so that to each group an element of the level above send its orders or signals. A very pertinent question to ask at the present time is the following: *why* is it so? *Why*, when a number of individuals are collected, they, beyond a certain number, do not form any more a structure capable of intelligent action, but just a crowd? Why does hierarchy appear to be an essential condition for life to be present?

I shall try to discuss, although very briefly, this question; in so doing, I hope that the methodology of approach and the scope and perspectives of cybernetics may be clarified. I hope also to show that, whereas in the past cybernetics, like all other sciences, has been heavily dependent upon the extant sciences, there is a fair chance that the situation may be in some cases reversed; that is, concepts and mathematical technologies may be developed in its study that may be of use to mathematics itself, logic and physics.

To clear the way from prejudice, some remarks are perhaps relevant. The first of them is that mathematics, as we have it today, is not an absolute necessity of nature, but just an historic accident, no more and no less than the fact that we happen to speak English, Italian or Russian; other forms of mathematics might well have been conceived (I am quoting von Neumann). The mathematics we have was invented mostly to study gravitation, and was later generalized to other phenomena of the physical world; the logic we use stems directly from Aristotelian logic, and is not certainly concerned with the detailed neurophysiological processes that go into the formation of thought. Very different forms of logic and of mathematics are indeed conceivable; one among them is the logic whereby nervous cells work; however different, they may lead, operationally speaking, to the same results. I offer one trivial example: suppose we are given a bow and an arrow, on a rainy and windy day, and have to hit a running deer. This is a classical problem of mechanics on which you can use electronic computers and analytic dynamics. I am only asking whether you would trust more, if you really need food, differential equations and computers, or a cannibal who cannot count up to three. The problem is not changed, whether computers or the cannibal solve it, but the techniques used in each case are certainly worlds apart. What are those used by a nervous system? This is one of the main questions cybernetics purports to study, and it again immediately leads to hierarchically organized systems.

Another warning whose necessity becomes immediately apparent when one is involved in studies of this nature, is that the cybernetic description of an "intelligent" system through hierarchies of levels does have profound implications. The function of an element at a given level need not be assumed to be more complicated than the function of an element at a different level; they only act each upon a different set of objects: think of a captain, or a general, or a soldier; they each have a different Weltanschauung or logic-this is well known-and a different domain under them. A simple example is that of a nuclear physicist and of a carpenter. The logic-if I may call it so-of the carpenter has absolutely nothing to learn from the logic of the nuclear physicist or from his knowledge, nor the other way around. The point is, that the age-long dream, which I myself nourished when a young student, that if one could know some fundamental equations everything would come out of them automatically on the basis of sheer mathematics, has to be completely forgotten. Now I deny any belief in that dream. As a human being, I belong to a given level, that is my logic. I can have not hope, no expectation of entering the logic of a higher or a lower level. My description of the world is confined to the level to which I belong.

Also very important is the fact that we should carefully distinguish between _laws_ and _rules_. No one would expect to find, by simply studying the equations of Maxwell and Lorentz, any notion on how to build a television or radio set; those are _machines_, they are built with a _purpose_, and to do so we must find rules. The same is true for any large system, such as a society or a language; _laws_ are required to know the general interaction patterns of the constituent elements (if we talk about the neurons of a nervous system, we attempt accordingly to find approximate equations to describe the mutual interactions of the firing patterns of those neurons); _rules_ must tell about the structure that a particular society, or army, or brain needs in order to accomplish most effectively its purpose.

The most important philosophical and

mathematical question in this domain is "why" are levels formed. Attempts do exist at describing hierarchical systems, once their structure is already given; what nobody has any idea of as yet, is "what" causes the formation of hierarchies. There must be some principle of economy, that is to say of maximum or minimum, which is satisfied only if such a structure is formed. At the end of this talk, I shall give intuitive evidence that it is so, in a way which I hope may soon become also a quantitative statement.

Natural Languages in written form (I should rather specify, in typewritten form) are ideal experimental material for trying to understand facts, rules and laws about very complex hierarchical systems. The statement is self-evident and needs no comment; the trouble is rather that language is far too difficult an object of study for present-day science. We must somehow resort either to simplifications of the real thing - and this I do not wish to do, because in any such simplification there is the overt danger of destroying just those features in which we are interested and in ending with a computer-language rather than a natural-one - or we must content ourselves with studying "simple things" at first. The latter is what Galileo teaches us to do, as he was the first to show that science is a profession of humility and not of pride, and that before talking about higher mechanics one should first know how stones fall.

Our approach to linguistic studies is based just on such premises; it aims at no more than finding the structural laws whereby individual letters are grouped into syllables, syllables into larger units, these into words, words into sentences and so on. Even with such simplifications, because we have killed any ambition to deal directly with questions like semantics or ambiguity (both much discussed and completely misunderstood subjects), the task which faces us is of extreme difficulty. We work with typewritten texts, so as to have a discrete rather than a continuous flow of symbols. Our approach is entirely phenomenological; unlike what several linguists do when they consider the particular language in which they happen to be interested as one among a host of possible ones, each being defined as a sequence of strings or letters, we take the language which we want to study as our whole world; in the same fashion, a physicist is not concerned with finding laws for all conceivable worlds, but only for the particular one in which he happens to live. Thus, when we study Italian, that is our world; we may later study English or French or Russian, and those again will be different worlds; a comparison between the structures of these different worlds might hopefully lead to common features. By reaching ever higher levels of abstraction one may hope to obtain, in the future, rules and laws valid not only for one language, but for most or perhaps all of them; since hope is the primary source of inspiration to any scientific investigation, also perhaps for languages such as the pictorial or musical, which are far removed from the customary written languages. Rather than expanding on generalities concerning things not yet discovered, let me only say that this approach leads to questions which are susceptible of precise mathematical formulation, or, when not yet so, pose problems in a form fit for quantitative speculation. It should also be added that it is the common consensus of experts in linguistics that the same questions which one poses when studying letters, syllables and so on, are found again at all higher levels of the structure of the language; thus to ask questions of a logical, of a syntactical, of a grammatical nature are things quite comparable in kind with those which form the present object of our study.

By language I shall mean from now on the Italian language; more specifically, some given text written without errors, letter by letter, word by word, in the memory of a computer: in short, the "text". "Language" is too wide a term, not easily definable and also changeable at will by a naughty experimenter who may invent words just to make trouble.

Suppose now that we have no a priori knowledge of the Italian language; for us, the text is only a collection of strings of symbols, the letters. On examining the text we find out, rather soon, that all the strings are written by using only 21 letters, this being the number of letters of the Italian alphabet. If we give one of the monkeys of which Bertrand Russel speaks a typewriter having a keyboard with only those 21 letters (plus a blank), the monkey will produce a text which has one, and only one, similarity with our Italian text: it is written with the same symbols. Otherwise, it is totally random. The text written by the monkey is called the <u>free monoid</u> generated by the "generating alphabet" constituted by the 21 Italian letters.

Make now a jump ahead and suppose that we have already solved our problem of finding out which are the Italian "syllables", that is the first buildingblocks of the language above the letteral level. Suppose now we give another monkey a larger typewriter on whose keyboard all of the extant Italian syllables, plus of course the blank, are reported, and let it also produce a text. If each syllable is represented not as a string of letters but as a symbol, such as a Chinese ideogram, the text produced by the second monkey is also a free monoid, whose generating alphabet is the set of the symbols which are representatives of the Italian syllables; it makes no difference that the syllabic alphabet has many more symbols than the letteral one.

Change now each syllabic symbol with the appropriate string of letters used in Italian to write that syllable; we obtain of course again a sequence of strings of letters, which evidently is only a part of the free monoid generated by the first monkey from Italian letters, a sub-monoid of it; it is also evident that the second monkey produces in some way a closer approximation to the Italian text than the first. The question we want to pose is the following: if we do not know a priori the generating syllables, can we reconstruct them from an examination only of the sub-monoid just mentioned? If this is possible we can retrieve all Italian syllables from a text in which no dividing line is set between syllable and syllable: a major achievement, one might say, in pattern recognition.

We find already in this very brief exposition of our problem its main characteristic: we have stated mathematically what we are looking for. What we have said holds only for infinite monoids and sub-monoids, however, with the further assumption that the Italian syllables be put on tape by a monkey, that is at random, which is clearly untrue of a real-language text. In order to proceed with the methodology of the physical sciences we must next try to separate the essential from the accessory. We can regard as accessory, for instance, the fact that any text we can actually analyze is a finite and not an infinite sequence of strings, as a monoid is supposed to be, nor can we ignore the fact that in a text in the Italian language syllables do not follow one another at random. We are thus left with a question which does not answer the one we find in reality, but comes as close to it as it is possible to ask from a physical model of the real world; supposing that we are allowed to neglect, because of our skill in handling the situation or because we decide to consider them only as errors, the finiteness of the text and the fact its syllables are not quite at random, how can we derive from it those strings of letters which constitute the Italian syllables, and only them?

It happens that the Italian language, as well as many others, is so written that the segmentation of the text into syllables is possible in a very neat way, by means of a technique that takes into account the fact that some letters are followed by only a very few, say for instance "q" only by "u", others, such as "a", are followed by many or all of the others. This technique can be proved mathematically, because it falls in line with a classic theorem of coding theory, to yield exactly the wanted result, provided the syllables be made in such a way that whatever is left of a syllable by cancelling its first letters is still an allowed syllable: thus if _stra_ is a syllable, so must be _tra_, _ra_, _a_. We have called this a _closed code_; after discovering this property from a mathematical analysis, it was soon apparent that it is indeed an obvious and desirable one in a natural language: it only means that if I can pronounce "stra", so can I pronounce "tra", "ra", "a".

Much work is still ahead, of course, because

we must now find ways whereby both the finiteness of the text and the influence of higher structural levels be neutralized as far as possible. I shall only say here that progress seems quite satisfactory, to the point of being total with the Italian language, fairly good with several others, such as Spanish, French and Malayan; we hope to be able to obtain, in a finite time, objective criteria to handle realistic texts in many more languages. Important features, that are certainly of marked interest for linguistics but may well have a deeper meaning for the whole subject of hierarchical systems, are showing up experimentally. It appears, for instance, that each language can be characterized with a very few numerical parameters, in a way which is of course gross yet sufficient for a machine to tell whether it is treating an Italian, a Spanish or a Polish text; this analysis is in principle _not_ a statistical one and does not make explicit use of frequencies; it bears no relation to Zipf's law, which applies to languages but also to sizes of towns, telephone numbers and everything else.

I have mentioned, avoiding technicalities, what we call the Procrustes program ("Procrustes" being an appropriate denomination for it, both because we use a cutting procedure to determine syllables, and because there is no doubt as to the fact that Procrustes is the forefather of all theorists). Among the features of this approach is the fact that the search for syllables is self-terminating: when all syllables are found, they are just printed out and that is the end of it. The program lends itself quite naturally to iteration, and in this sense we can think of a "vertical Procrustes program", in which at the first level we find letters, at the second syllables, at the third words (so to say), etc. A host of new, fascinating problems arise, though, if we consider that the whole thing becomes soon unwieldy: starting from 21 letters in Italian we find, more or less, 600 syllables; iterations would yield increasingly larger numbers. We need therefore something else; a technique whereby the elements of a given level may be collected into classes, each of which may be considered as an individual element for the purpose of further iteration; ordinary grammar does this when it says that "cat", "dog" and "chair" are "nouns", and likewise with adjectives, verbs and other parts of speech: this kind of intuition is absolutely precious, and could not be had if we were handling other systems equally complex but of a different, say biological, nature. We work thus on what we may call the "horizontal Procrustes program"; although the work has barely begun, the indications we find are stimulating: problems have a way of proposing themselves to our attention, most being completely unsuspected by us before they arise as logical necessities.

I shall take now the liberty of speaking of things not yet done, but for which we can see some plan of research, in order to give a perspective on what, with luck, we might expect to obtain from such investigations. Suppose we consider only, to fix ideas, sequences of two letters, that is "digrams"; to indicate them in a form convenient for mathematical computation, we write a matrix in which there are 21 rows, each dedicated to the end letter of the digram, and 21 columns each dedicated to its first letter. Thus we can put "1" wherever there exists a digram at the crossing of a column with a row, "0" otherwise: we have an "incidence matrix" (made i.e. of "0" and "1"). A convenient definition of class of digrams, etc., obtains if we decide that a class is formed by all, and only, elements "1" which are aligned either horizontally or vertically with one another; in other words, suppose we have a full rectangle of "1" in that matrix, it denotes a class, even if its rows and columns are pulled apart and lines of "0" interposed (an obvious advantage of this definition is that it stays invariant under permutations of rows and columns, which is important because names of letters must be immaterial if we are interested in their relations). A heuristic procedure was found to deduce about 130 distinct classes out of about 600 syllables arranged in this way. We decided then to suppress the condition that classes should be disjoint from one another; that is, we allowed a syllable to belong to more than one class. As soon as this was done, the number of classes collapsed to about 25, clearly of the same order of magnitude as the letteral alphabet. It became also apparent that a more efficient way of memorizing syllables is to do so separately for the component letters and for the order in which they are written; this agrees well with what psycophysics tells about the relative ease of memorizing a string of unrelated words with, and without order. At this point we notice that the structure of the Italian language is such, that there are syllables which belong to one, others which belong to two, others to more classes; this is clearly a "semantic" information, quantitatively stated and related to structural properties. A quantitative definition of semantics defies yet all speculations; it is therefore promising, in our opinion, that such a notion should arise so naturally, without our asking for it. One should of course expect a host of measurable "semantic" quantities along the way, there being absolutely no reason why semantics should mean one quantity, and not a whole category of them.

Another fact immediately also called upon our attention. I consider it quite important, and although we have done no work yet on it, it may be one of the most significant outcomes of this investigation. If we want to learn a language we have to learn everything about its rules, that is about its classes and sub-classes, levels and sub-levels, and so on. Suppose now, on a simple example again for the sake of clarity, that we wish to memorize a class which is represented by a rectangle which is entirely full (of "1"); to do so, since we have already memorized that we are only interested in rectangles, we have to put into our storage only the two coordinates of the lower left and of the upper right corner. Suppose now that the situation is slightly changed; that we have, for instance, the same rectangle as before but with two holes (that is "0") into it; we want now, again, to memorize the "rules", that is the classes. Since rules mean "rectangles", it is clear that we have to split the rectangle into sub-rectangles, so as to have them all; this may mean quite a lot of coordinates to memorize. We can proceed in a different way, though, by first memorizing the whole rectangle as if it were full, without holes; then, apart, we can memorize the locations of the holes, considering them as "exceptions". We shall have memorized in this way one rule plus some exceptions. If you just think it over for a second, you will find that the second procedure is by far more economical than the first one.

It also appears to be the one which is naturally followed by the human mind; we are used to syllogistic reasoning as the foundation of mathematical logic; thus, "humans are mortal, Socrates is human, therefore Socrates is mortal". But the way the human really thinks is not certainly this; everybody's experience will confirm that the way our mind works is the following: "Joe belongs to the Mafia, Joe is Italian, therefore all Italians belong to the Mafia". This is of course inductive versus deductive reasoning; inductive reasoning is a much more complex subject of investigation, which in our perspective of the world can only proceed by trial and error; first by taking something as generally true, then finding that it is not always true; at this stage we make up exceptions to account for that, and when exceptions become too many we readjust our set of rules. And so on. It is a statement which needs no proof to say that logic based upon such notions has little to do with the formal logic which most people consider at present as the only conceivable form of it. I wish only to add that all other studies made on the working of nervous systems by means of mathematical models lead exactly to the same conclusions; should we wish to build some machine to do the kind of linguistic analysis I have been mentioning thus far, it would most naturally be one that works according to the general principles of our theory of neural nets.

In conclusion, I may be allowed to add some guesses on the mechanism of level formation in a hierarchical system; as I have said at the beginning, this is the crucial point in the study of complex systems. I wish to suggest that level formation occurs only because it is more economical, in all possible meanings, for a system to organize itself hierarchically. I can offer no proof, but an intuitive example may carry perhaps conviction to a sympathetic listener.

I shall consider only a seemingly trivial question. When we play poker or some other game which involves transactions of money, we use chips as tokens for money. We do not use, however, chips of only one denomination, say "white" meaning "one dollar" as the unit, but chips of different colors, each color denoting some larger multiple of the basic unit. Suppose now that we have some gain function which tells how much we gain by increasing the number of the elements; typically, this function may be an entropy, and we may report it on the vertical axis, with the number of states on the horizontal axis; when the number of states increases we expect of course a logarithmic curve. This means that if the number of states increases by a few units, starting from the origin, the gain is nearly linear, but soon later we have to add a large number of units again to obtain a comparable gain.

In our case, talking about chips, it is clear that from the point of view of the poker player (or if they are soldiers, from the point of view of the commanding officers) computations must be made with Bose rather than Boltzmann statistics: we are not interested in "which" white or red chips we have, but only in "how many", the identity of chips being totally irrelevant. It is also clear that if we keep only white chips, each worth the lowest allowed betting, we can bet any amount; however, this might be inconvenient. The introduction of chips of higher denomination means that we have taken a "clustering" of white chips and substituted the whole of them with a red chip; red chips being each worth, say, ten of the white ones. If we think of the entropy curve I was mentioning before, we see immediately that the clustering of a number of elements into one of different species, followed by similar iterations of the procedure, changes the former smooth logarithmic plot into another plot, which at each change of denomination, at each new clustering that is, exhibits a discontinuity in the derivative. We end up with a new curve; the first arc coincides with the first arc of the original curve, then a steeper arc comes because an arc of the first curve has been contracted and scaled up, then again a steeper one, and so on. We have introduced a curve composed of a sequel of logarithmic arcs, each of which differs by a scale factor. The result will clearly be something which may approximate the behavior of a straight line, that is of a linear gain, much better than we might ever obtain otherwise.

It is therefore at least intuitive, and probably not difficult to change into a quantitative, formal theory, that level formation occurs in terms which can be made mathematically precise by specifying that some variational principle has to be satisfied. It does not take much of a stretch of imagination to realize that each time a new level is formed, or there is a transition in between levels, we are confronting a situation which, if we were doing physics, we would call a phase-transition. This last remark justifies, I hope, my initial remark that a study of cybernetics may prove useful to the physical sciences, as well of course as to the others concerned with large complex systems, among which mathematical economics is certainly becoming of major importance.

LIST OF PARTICIPANTS

Edward Ames
Department of Economics
State University of New York at Stony Brook

Joseph Aschheim
Department of Economics
George Washington University

M. A. B. Beg
Department of Physics
Rockefeller University

Robert Blumenthal
Department of Health, Education and Welfare
National Institutes of Health

Martin Bronfenbrenner
Department of Economics
Duke University

E. R. Caianiello
Consiglio Nazionale delle Ricerche
Laboratorio di Cibernetica

Mou-Shan Chen
Center for Theoretical Studies
University of Miami

Anthony Colleraine
Department of Physics
Florida State University

P. A. M. Dirac
Department of Physics
Florida State University

John Eccles
Department of Physiology
State University of New York at Buffalo

Erich A. Farber
Department of Mechanical Engineering
University of Florida

Sidney Fox
Institute for Molecular and Cellular Evolution
University of Miami

Nicholas Georgescu-Roegen
Department of Economics
Vanderbilt University

Donald A. Glaser
Department of Molecular Biology
University of California at Berkeley

Melvin Gottlieb
Plasma Physics Laboratory
Princeton University

Gary Higgins
Lawrence Livermore Laboratory
University of California

Joseph Hubbard
Center for Theoretical Studies
University of Miami

C. S. Hui
Center for Theoretical Studies
University of Miami

Henry Hurwitz
General Electric Company
Schenectady, New York

Abraham Klein
Department of Physics
University of Pennsylvania

Behram Kursunoglu
Center for Theoretical Studies
University of Miami

Willis E. Lamb, Jr.
Physics Department
Yale University

Joseph Lannutti
Department of Physics
Florida State University

Sydney Meshkov
Radiation Theory Section
National Bureau of Standards

Stephan L. Mintz
Center for Theoretical Studies
University of Miami

Laurence Mittag
Center for Theoretical Studies
University of Miami

Lars Onsager
Center for Theoretical Studies
University of Miami

Edwin E. Salpeter
Laboratory of Nuclear Studies
Cornell University

Julian Schwinger
Department of Physics
University of California at Los Angeles

George Soukup
Center for Theoretical Studies
University of Miami

Ichiji Tasaki
National Institute of Mental Health
Laboratory of Neurobiology

Edward Teller
Lawrence Berkeley Laboratory
University of California

Georges Ungar
Department of Anesthesiology and Pharmacology
Baylor College of Medicine

Jan Peter Wogart
Institute of Inter-American Studies of the Center for Advanced International Studies
University of Miami

OBSERVERS:

George Adelman
Managing Editor
and Librarian
Neurosciences Research
Program
Massachusetts Institute
of Technology

Robert Lind
Department of Physics
Florida State University

F. David Peat
National Research Council
of Canada

SUBJECT INDEX